AF270242

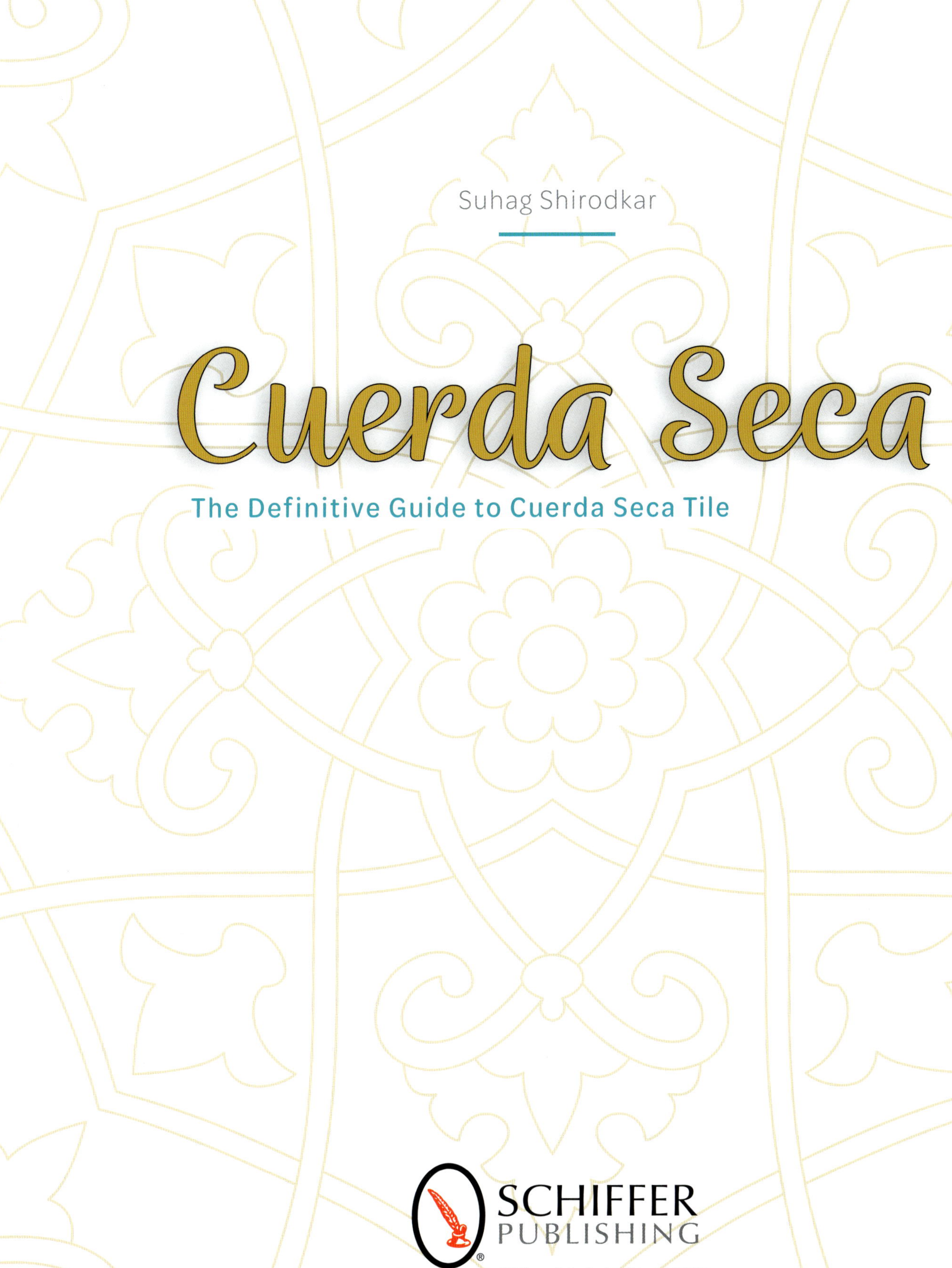

Suhag Shirodkar

Cuerda Seca

The Definitive Guide to Cuerda Seca Tile

SCHIFFER PUBLISHING

4880 Lower Valley Road • Atglen, PA 19310

Copyright © 2022 by Suhag Shirodkar

Library of Congress Control Number: 2021942415

All rights reserved. No part of this work may be reproduced or used in any form or by any means—graphic, electronic, or mechanical, including photocopying or information storage and retrieval systems—without written permission from the publisher.

The scanning, uploading, and distribution of this book or any part thereof via the Internet or any other means without the permission of the publisher is illegal and punishable by law. Please purchase only authorized editions and do not participate in or encourage the electronic piracy of copyrighted materials.

"Schiffer," "Schiffer Publishing, Ltd.," and the pen and inkwell logo are registered trademarks of Schiffer Publishing, Ltd.

Edited by Karla Rosenbusch
Designed by Danielle D. Farmer
Cover design by Danielle D. Farmer
Type set in Wreath / Omnes / Omnes Cond

ISBN: 978-0-7643-6309-2
Printed in India

Published by Schiffer Publishing, Ltd.
4880 Lower Valley Road
Atglen, PA 19310
Phone: (610) 593-1777; Fax: (610) 593-2002
Email: Info@schifferbooks.com
Web: www.schifferbooks.com

For our complete selection of fine books on this and related subjects, please visit our website at www.schifferbooks.com. You may also write for a free catalog.

Schiffer Publishing's titles are available at special discounts for bulk purchases for sales promotions or premiums. Special editions, including personalized covers, corporate imprints, and excerpts, can be created in large quantities for special needs. For more information, contact the publisher.

We are always looking for people to write books on new and related subjects. If you have an idea for a book, please contact us at proposals@schifferbooks.com.

Other Schiffer Books on Related Subjects:
The Ceramics Studio Guide: What Potters Should Know, Jeff Zamek,
 ISBN 978-0-7643-5648-3
What Makes a Potter: Functional Pottery in America Today, Janet Koplos,
ISBN 978-0-7643-5811-1
20th Century Decorative British Tiles: Craft & Studio Tile Makers, Chris Blanchett,
ISBN 978-0-7643-2468-0

CONTENTS

FOREWORD

Cuerda seca lunette from the fifteenth century Yeşil Türbe mausoleum of Ottoman Sultan Mehmed I, Bursa, Turkey.
Patricia Blessing

What is it about tiles? It is an evocative medium. Their appeal spans the globe, and part of the intrigue evolves from their antiquity—the history of tile making and crafted glazed tiles can be traced back over 5,000 years.

Human hands take moist clays from the earth and push, squeeze, press, and mold this malleable material, and when it is fully formed, dried, glazed, fired, and installed in place—it is a miracle!

There are many methods for designing and creating beautiful, colorful tiles. This book in your hands addresses the cuerda seca (dry line) tradition. This method, as you will read, was developed in the fourteenth century. By defining and outlining patterns on the tiles that allowed all the bright colors the artisan wanted to use to remain separate, beautiful polychrome tiles with well-defined boundaries between the colors were readily produced.

However, as you will see our author point out, "It's not easy to make good cuerda seca tiles"; hence the journey she embarked on that gave birth to this book. We have interfaced many times throughout this expedition. No "particle of clay" has been left unturned in the book's production. *Cuerda Seca: The Definitive Guide to Cuerda Seca Tile* covers every possible question you may have, from a discussion on the origins of cuerda seca and its definition, to the best practical studio practices you may want to exercise to have a successful outcome for your bright and beautiful tiles.

Suhag writes with sublime technical clarity; her research and quoted resources are impeccable. The resources are unique, too, since the author lives and works in Goa, on the west coast of India, as well as sometimes in Northern California. There has not been a definitive how-to manual in the past dedicated to making cuerda seca tile. The author spends an entire chapter assisting the reader on how to use this inspiring, well-organized book, which is also filled with fine images of cuerda seca tiles in situ both from historical and modern sites.

So, dear reader, fire up your imagination—there is no limit to what you may accomplish in your own studio, perfecting beautiful tiles by using this definitive cuerda seca guide!

Sheila Menzies, executive director
Tile Heritage Foundation
www.tileheritage.org

My cuerda seca journey began at the dawn of the internet. Although we had knowledgeable and sharing people on ceramics listservs back then, information was hard to find and corroborate. I had spent fruitless weeks testing formulas for cuerda seca's distinctive black line. Late one summer evening, frustrated with my results, I called Richard Keit of RTK Studios in Ojai, California. Although I barely knew him, I presumptuously asked for potentially proprietary information. I can still hear Richard's words: "It's just manganese oxide and linseed oil, isn't it?" This book might never have happened but for that generous response.

Sheila Menzies and Joe Taylor at the Tile Heritage Foundation in Healdsburg, California, responded to my questions with warmth and enthusiasm. If they didn't have an answer, they pointed me to someone who did, and followed up until I got the necessary information.

Setting up my studio in Goa, India, meant a whole new set of challenges, from rewiring my Skutt kiln for the domestic supply to testing local clay and glaze ingredients. Sandeep Manchekar of Anvi Pottery shared knowledge and resources freely. His cheerful attitude made the transition a joyous adventure in ceramics rather than a chore.

In Bangalore, Leila Bose Powar was an outstanding mentor. Firings of her gas kiln meant sumptuous meals and engaging conversations on the arts, punctuated with vignettes from her adventurous life. Although Leila is no more, her commitment to her craft continues to inspire.

Anand, Mohini, and Mallika kept our little homeschool going while I was away firing. Their superior design sense continues to save me from gaffes. My best work is for them and because of them.

A good editor pushes a book up several notches. Thanks to Sandra Korinchak and Karla Rosenbusch at Schiffer Publishing.

And thanks to you, dear reader. You will take this craft further ahead. Please send in your comments on using this book to www.flameback-studio.com, so that we can improve future editions. Together, we will draw even more artists and connoisseurs into the enticing world of cuerda seca tile.

The Cuerda Seca Tradition

The glazing technique called *cuerda seca*, in which brilliant glazes are separated by lines of dark resist, is thought to have originated in central Asia in the fourteenth century. Previously, craftsmen wanting to make vibrant ceramic panels with sharp color boundaries had to use mosaic techniques, laboriously cutting tiny pieces of monochrome tiles and assembling them into patterns.

Cuerda seca made the job faster and easier. Craftsmen could now make multicolored tiles with the visual impact of mosaic, every color cleanly and clearly demarcated from every adjacent color. Entire facades, gateways, and mausoleums could be clad in brilliant color with a speed and ease not possible before.

Lion on a shield, Spain, late fourteenth to early fifteenth century. *MetMuseum.org*

Opposite: Cuerda seca facade of the fourteenth-century Aq Saray palace in Shahr-i-Sabz, Uzbekistan. *Patricia Blessing*

CHAPTER 1

Top: Reciting Poetry in a Garden, cuerda seca panel, Iran, seventeenth century. *MetMuseum.org*

Bottom: Gathering in a Garden, cuerda seca panel, Iran, seventeenth century. *MetMuseum.org*

Opposite page

Top: Peacock fountain at the Adamson House in Malibu epitomizes California's love affair with cuerda seca tile. *Diana Mausser*

Bottom: Detail from the Adamson House fountain. *Diana Mausser*

Good ideas travel fast. Across Iran, Turkey, and Spain, cuerda seca became an important technique in the ceramic artisan's repertoire, along with mosaic and various underglaze and overglaze methods. In the centuries to follow, it enjoyed waves of popularity across many regions of the medieval world, all the way east to India.

Fast forward to the early twentieth century, when a new style of architecture called Spanish Colonial Revival emerged in California, drawing inspiration from Moorish, Persian, and Turkish design. Now cuerda seca, along with its sister technique *cuenca*, enjoyed a great flowering. Its bold outlines and vibrant color palette were ideal for conveying the new California style. Several potteries were established along the Pacific coast to cater to the burgeoning demand. Cuerda seca tile lavishly embellished haciendas and resorts, its exuberance evoking a sun-kissed lifestyle of wealth and leisure.

Modern cuerda seca artists have access to materials and equipment like never before. We enjoy a sumptuous palette of glazes. We can control our processes to achieve consistent line and color, firing after firing. In our well-lit and well-equipped studios, we can stand on the shoulders of artisans of the past seven hundred years, keeping the cuerda seca tradition vibrant and alive into the future.

Tile believed to be from the tomb of
Rumi, Konya, Turkey, fourteenth century.
MetMuseum.org

CHAPTER 2

The Defining Black Line

We do not know what the earliest cuerda seca artists called their technique. But by the sixteenth century, the Spanish term "cuerda seca" was used for a technique to separate glazes with a dark, oily resist line.

In scholarly literature, cuerda seca is most often translated as "dry cord," a reference to the appearance of the resist line after firing. Some scholars suggest that a more idiomatic translation is "colorless line." However, unlike a wax resist line, which burns off in the kiln to reveal the underlying clay, the cuerda seca line persists after firing and contributes to the aesthetic of the finished tile.

While we may debate the terminology, few would dispute cuerda seca's unique appeal. Because the oily resist line pushes back aqueous glaze as it is applied, glaze builds up between lines. It is this distinctive pooling of glaze within resist lines that gives cuerda seca tile its dimensionality and tactile allure.

Freshly applied glaze pools
within resist outlines

Cuerda seca works best on surfaces that are fired flat, including tiles. You could use the technique on plates and platters so long as their curvature is gentle. On vertical surfaces, there is a risk of glaze dripping down across resist lines during firing.

How to Use This Book

Until now, there was no how-to book dedicated to making cuerda seca tile. I suspect most cuerda seca artists taught themselves as I did—through trial and error, with more error than they desired. Let's hope this book changes things. It's written primarily for ceramic artists who wish to add cuerda seca to their repertoire. It is also for cuerda seca aficionados, so that they can appreciate this remarkable technique.

It is not easy to make good cuerda seca tile. You need a suitable clay body, and glazes that fit. You must select your design carefully and match your glaze palette to the design. You must visualize where your tile will be installed and how it will be handled. This makes you a ceramist, a designer, and an architect rolled into one.

With this book as your companion, you will tackle this daunting process a step at a time, setting you on the road to success.

Let's begin by thinking about the tile. Any glazed tile consists of two layers: a thick layer of clay coated with a much thinner layer of glaze. The clay should be strong and mature after bisque firing yet remain semivitreous, to ensure a permanent bond at the clay-glaze interface. It must meet several other requirements, including low shrinkage and near-zero warping. Section 2 covers clay body formulation, and section 3 describes forming and firing flat tile.

CHAPTER 3

Should a cuerda seca design be traditional or modern? How do we assess whether a design can be married to cuerda seca's bold glaze palette? How can we create murals that engage and delight viewers? Design principles and techniques for translating patterns into cuerda seca are discussed in section 4. Design inspiration is available throughout the book, most directly in section 9. New vector patterns for cuerda seca will periodically be uploaded to www.flameback-studio.com.

Section 5 focuses on the cuerda seca resist line—how to formulate the resist and apply it on bisque tile.

Cuerda seca requires glazes that mature in the earthenware range, between firing cones 06 and 04 (980°C–1,040°C or 1,800°F–1,915°F). Glazes and glaze firing are covered in section 6, while section 7 covers installation of the finished tile on walls and furniture.

The "Resources" section tilts slightly toward India and California, the two places where I make cuerda seca tile. I have listed general and online sources wherever possible and encourage you to write in with resources for your geographic region.

Throughout the book, you will find troubleshooting tips. I have shared my missteps wherever I thought they might serve as beacons for you. The goal is to make your cuerda seca journey smoother than mine.

Opposite page

Photo: Diana Mausser

SECTION 1

Defining Studio Space

To make cuerda seca tile, you need well-lit, well-ventilated studio space, with suitable plumbing and code-compliant electrics. Since you will handle clays, glazes, and oxides, the studio should be a dedicated space that is not part of your living area. A studio layout that prioritizes workflow and ergonomics creates an efficient and pleasant environment for producing high-quality tile.

The gray glaze has crawled. Glaze may crawl if dust settles on a tile before it is fired.

Space Requirements

The scale of your operation dictates how much space you need. Whether you plan to make cuerda seca tile occasionally, regularly, or as a studio production, try to dissect the process to determine space requirements at each step. Dust control is imperative from start to finish. Here, broadly, are the stepwise requirements.

STEP	POTENTIAL FOR CREATING AMBIENT DUST	REQUIREMENTS & CONSTRAINTS	IDEAL LOCATION
Processing clay (weighing, sieving, soaking, pugging, and packing into bags for later use)	High	Should be close to the raw-materials storage area, with easy access by vehicle or wheelbarrow.	If weather permits, an open-fronted shed or patio, with an attached storage room for raw materials. Alternately, a well-lit and well-ventilated indoor space.
Forming and drying tile	Medium	Well-lit and well-ventilated space that permits slow, even drying of green tiles. Good airflow is essential. No direct sunlight on the tiles, since this will cause uneven drying and warping.	Indoor space, with grid racks for drying tile
Preparing glazes	Medium	Glaze mixing should not share space with clay processing.	Well-ventilated indoor space.
Firing (Electric)	Low	Dedicated electrical supply that meets safety standards. Open space and ventilation around the kiln as per the manufacturer's guidelines. Easy loading and unloading.	Indoor space with (1) grounded electrical outlets meeting specifications, (2) kiln vents, if the space is not adequately ventilated, and (3) shelves for kiln furniture and for loading and unloading tile.
Application of cuerda seca lines and glazing	Negligible	Dust should not settle on tile surfaces. Good natural light on work benches. Easy access to glazing tools and materials. Brushes and screens must be cleaned outdoors to minimize solvent vapors.	Airy, well-lit "clean room" with shelves and easy access to tools and materials. Outdoor space, such as a patio, for cleaning and drying screens.

Storage

A studio is efficient and pleasant when storage is assigned up front, not as an afterthought. Here are storage requirements for a cuerda seca studio.

MATERIALS TO BE STORED	REQUIREMENTS AND CONSTRAINTS	IDEAL LOCATION
Heavy bags of clay raw materials	Should be close to the clay-processing area but away from anything else. Vehicle or wheelbarrow access for ease of unloading.	A dedicated storeroom or shed with open floor space, a few shelves, and room for a weighing scale.
Painting and screen-printing supplies and glaze ingredients	Dry, dust-free environment	A storage cabinet or bins easily accessible from the glazing station
Kiln shelves and furniture	Close to the kiln	Heavy-duty storage shelves in kiln room
Small tools and equipment	Closest to point of use	Multiple, small shelves above work benches and slab roller
Work in process	Separate shelving for green tiles and bisque tiles	For green tiles, open-grid racks to facilitate even drying. For bisque tiles, dust-free shelves.
Finished tiles and packing materials	Well-demarcated area, away from the primary process flow	Clean shelving with space for storing packing materials. Display racks for walk-through visitors and customers.

Glaze and resist materials neatly shelved for easy access

Studio Layout

Workflow

In a well-designed studio, processes that are sequential or have similar requirements stay close together. Materials and people move smoothly without crisscrossing. Studio layout "follows the clay," which means that the path from raw clay to green tile to bisque tile to glazed tile to finished tile tracks a smooth line or arc through the studio.

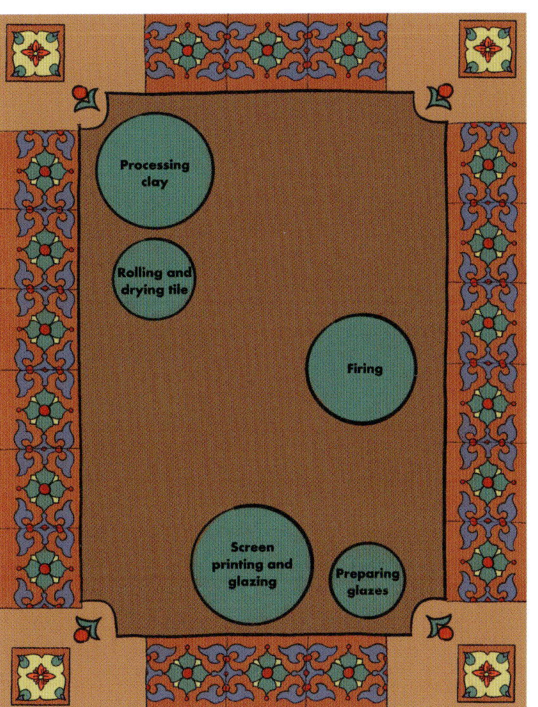

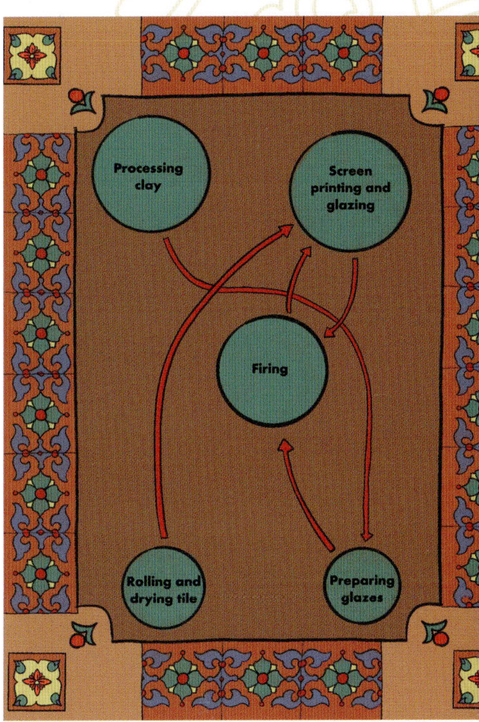

It is useful to think about studio layout conceptually before diving into the nuts-and-bolts placement of workbenches, sinks, or electrical outlets. For a rectangular studio space, for example, potential layouts could be like the two shown above at left and center. In both cases, the dust-producing processes lie close to each other and away from the cleaner post-bisque processes. The kiln is equally accessible for bisque and glaze firings.

At top right is a layout that will not work well. Although the dusty processes are still away from the clean, movement through the studio is inefficient, as the arrows reveal. This might be simply an annoyance if you work alone, but a serious impediment if you have coworkers or studio assistants.

Unless you are building a studio from scratch, it is unlikely that you will be able to configure it ideally. The locations of existing electrical outlets, drains, plumbing lines, and ventilation will certainly influence your layout. Nevertheless, conceptualizing a layout helps you optimize within the physical constraints and impediments.

Good layouts (left and center) and a poor layout (right) for a rectangular studio. *Mohini Bariya*

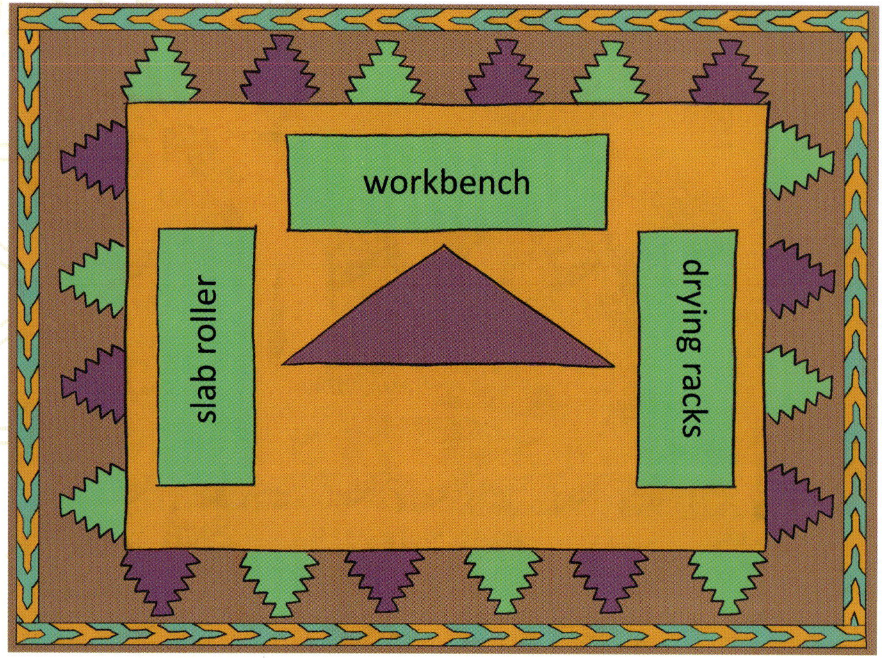

Work Triangles

The concept of a work triangle comes from kitchen ergonomics, in which a refrigerator, a sink, and a cooktop are visualized as three vertices of a triangle within which a cook operates. The idea is to accomplish tasks by reaching, stretching, and pivoting rather than walking up and down. This concept can be directly applied to a studio to increase efficiency and decrease fatigue. The process of rolling out and drying tile can, for example, fit into a work triangle. Distances between the vertices of a work triangle should be based on personal comfort.

Adjacent work triangles can be visualized for the screen-printing and glazing steps, with shared work-in-process shelves between the two stations.

Above: The work triangle, as applied to tile rolling and drying. *Mohini Bariya*

Right: Sequential processes, both requiring a dust-free environment, visualized as adjacent work triangles. *Mohini Bariya*

Ideal bench height depends on the task to be performed at that bench. *Mohini Bariya*

Bench Height

Ideal workbench height depends on the nature of the task performed at that bench. While you may not have the luxury to build a custom bench for every task, it helps to know what works best for each.

For tasks performed sitting down, such as glazing, the standard desk height, 29" (74 cm), suits most people. For tasks performed standing, such as rolling out, cutting, and finishing tile, the ideal bench height is 4" (10 cm) below the elbow. And finally, for standing tasks that require you to bear down, such as wedging clay, the ideal bench height is 8" (20 cm) below the elbow.

If you cannot dedicate a bench for wedging, consider building a stable, nonskid wooden step, 4" (10 cm) high, to stand on while you wedge.

Dust Control in the Studio

Dust is inevitable in a clay studio. Silica in clay dust damages lungs and can cause lung cancer. Dust can also settle on freshly glazed tiles, causing crawling or other unsightly effects.

You can minimize the hazards of silica dust by wearing a mask or respirator while weighing out dry clay or glazes. In addition, these studio practices will minimize ambient dust:

- Wipe down worktables frequently with a damp sponge.

- Wet-wipe all other work surfaces as well. Slab roller grooves trap clay dust and must be cleaned regularly.

- Keep mouths of clay bags tightly twisted and closed.

- Do not drop bags of clay from a height.

- Keep lids tightly shut on bins, boxes, or bottles containing dry materials.

- When weighing or sieving dry materials, do not tap or hit scoops, sieves, or the weigh scale.

- Immediately after wiring off a slab of clay, run fingers over the wire to remove residual clay before it dries on the wire.

- Finish green tiles while they are still leather hard. Finishing bone-dry tiles creates more dust.

- Use a lidded bin for clay scraps. Gather up and discard clay scraps into the bin frequently.

- Set up a weekly routine to wash sponges, work towels, and slab roller canvases. Do not allow clay to build up on canvases.

- Vacuum the kiln interior regularly.

- Clean up at the end of each work session, rather than postponing to the proverbial tomorrow.

Slab roller grooves are dust traps. Do not use a wire brush for cleaning, since it will wear the grooves down over time.

Equipment and Tools

There are many excellent reference books on equipping a ceramics studio (see "Resources"). This chapter discusses equipment and tools specific to cuerda seca tile.

Equipment

Slab Roller

If you plan to make tile regularly, invest in a dual-roller slab roller. Dual rollers equalize shear forces to the top and bottom of the wad of clay as it is pushed through, and therefore lower the chances of tile warping. A slab roller with a wagon wheel, rather than a crank handle, makes rolling easier. Gauges at either end of the rollers ensure that slabs are even.

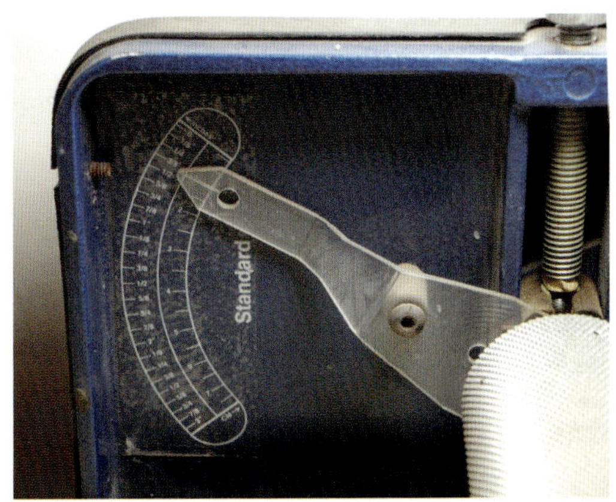

A gauge like this on both sides of the roller bed helps roll even slabs.

Slab roller with dual rollers and a wagon wheel. Location: Clay Planet, San Jose, California

Heavy-duty canvas works well with a slab roller and lasts a long time. Unless a design is very detailed, the texture of canvas on wet clay will not interfere with the design. If you want smoother tile, rib over the slab before cutting out tile. Alternately, use SlabMat sheets instead of canvas. SlabMat is a nonwoven fabric for rolling out texture-free slabs.

There are many other ways to make tile, without using a slab roller. You could pop tile out of plaster molds with a ram press for faster production. You could also extrude tiles, employing good equipment and techniques to minimize warping.

Clay Mixer and Pug Mill

A clay mixer mixes dry clay ingredients with water to form a mass of wet clay. After it is mixed, the wet mass must be manually moved to a pug mill, which maneuvers it into a uniform clay body, ready to wedge. Some of the more expensive pug mill models include a de-airing vacuum attachment to produce a uniform, de-aired clay body that requires little or no hand wedging.

Studio-size mixing pug mills, which combine the functions of mixing and pugging, are expensive but decrease the manual labor of hand mixing and

A de-airing mixer and pug mill take away most of the grunt work of processing clay. Claystation, Bangalore, India.

wedging clay. Some de-airing models come with optional dies for extruding tile. Budget permitting, look for a model that mixes, de-airs, and pugs. It should accept dry clay as well as reclaimed clay and be easy to disassemble for cleaning.

Kiln

Electric or gas-fired kilns can both be used for cuerda seca tile, although electric kilns yield more consistent color, both within a single firing and across multiple firings.

Track electricity consumption for every firing. A sudden drop in electricity consumption for a firing may indicate a damaged element, while a spike may point to inadvertent heat loss.

Kiln Furniture

Although you could fire tiles on standard kiln shelves, tile racks or rods increase efficiency. Because handmade tiles are not as thin and uniform as machine-made tiles, standard tile setters do not always provide sufficient headroom. Make sure the tile furniture you select provides the headroom required for handmade tile.

Kiln rods support tile for a glaze firing. The rod supports, which optimize kiln usage, were designed by Jim Sullivan and Diana Mausser of Native Tile. *Diana Mausser*

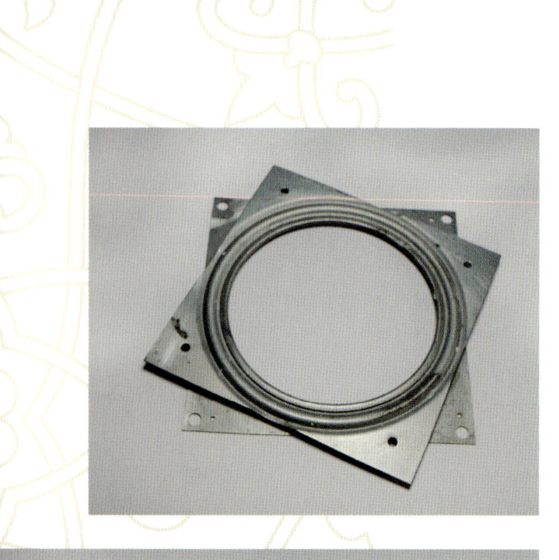

Top: Swivel plate for ease of glazing

Bottom: A thrust bearing like this one, glued under a piece of board, makes a hardy swivel plate.

Left: Spring-loaded stainless-steel tile cutter

Center: An "alternative" cutter

Right: Tile from a poorly designed cutter lacks a straight edge.

Tools

Swivel Plates

Tile placed on a swivel plate can be rotated while it is being glazed, making glazing faster and easier.

To make swivel plates in custom sizes, superglue a thrust bearing under a square plywood piece of the right size.

Although some cuerda seca artists use a banding wheel for glazing, I find that the rim of the banding wheel can rub against the wrist, causing mild abrasion and fatigue. This is especially true for detailed patterns.

Tile Cutters

Spring-loaded cutters are the workhorses of a tile studio. They are available in several shapes and sizes—squares, rectangles, circles, diamonds, and polygons. The spring helps release the tile slowly and evenly, minimizing warping of the wet tile. Dusting the cutter face with talc facilitates release.

Since heavy-duty spring-loaded steel cutters are expensive, it might be tempting to purchase a cheaper alternative. Here are some examples (below, center, and right) of issues with nonstandard cutters.

Notice that the cutter's plunger is lopsided and the circular opening much too large, which makes the plunger wobble as it is pushed down. Wobbling warps the wet tile. Notice also that the metal sheet does not quite meet at the corner. Clay accumulates in the gap, and the resulting tile will have a jagged corner.

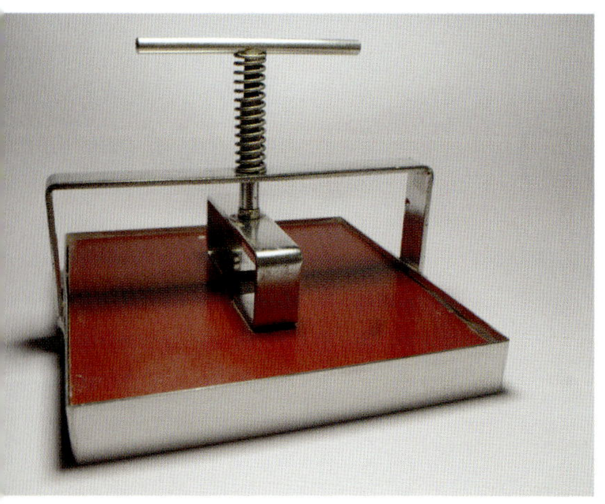

Left: Rubber bulbs, with precision tips for fine work

Below: Clear bottles with precision tips of various sizes

Bulbs and Precision Trailers

Rubber bulbs with precision nozzles, sometimes sold as slip trailers, work well for glazing cuerda seca. Their disadvantage is that you cannot see how much glaze remains inside the bulb and therefore the last few drops of glaze can sputter on the tile.

An alternative is these inexpensive clear bottles with precision tips. You can see how much glaze remains in the bottle and swap to finer tips for smaller glazing areas. They are sold online as "solder flux dispensers."

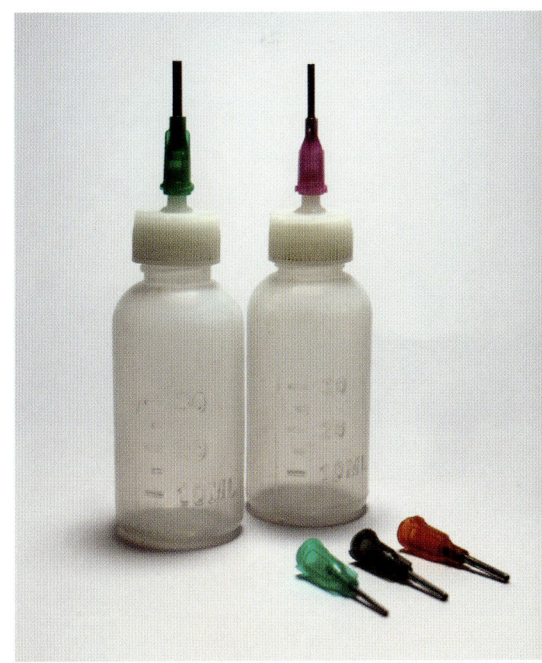

SECTION 2
THE CLAY BODY | 6: Clay Body Basics | 7: Formulating a Terracotta Tile Body | 8: Testing the Tile Body | 9: Fine-Tuning the Tile Body

Clay Body Basics

The strength and durability of tile comes from the composition of its clay body. In this chapter we review some basic concepts, our objective being to formulate a terracotta clay body for cuerda seca tile.

Formulating a clay body versus simply following a recipe helps you appreciate the contribution of each ingredient and make substitutions or trade-offs whenever necessary. Equipped with this understanding, you can develop a clay body for your specific requirements and make informed choices about manufactured bodies. If you plan to use manufactured bisque, you can ask the supplier about its clay body and anticipate how it might respond to your glazes and firing sequences. Paying attention to your clay body, whether prepared or purchased, enables you to make tile of lasting strength, utility, and grace.

Clay or Clay Body?

Clay is the sticky, slippery stuff that charmed us into becoming ceramic artists. Somewhere on this planet there might lie a magical seam of clay that can straightaway be fashioned into strong, flat tile. In reality, tile made purely from a native clay shrinks unevenly, warps as it dries, or cracks in the kiln. The clay needs to be transformed into a clay body.

Earthenware clay bodies, pugged, de-aired, and ready for use. Location: Clay Planet, San Jose, California.

Opposite page

A kaolin mine in Gujarat, western India

Top: Seams of clay intersperse with other sedimentary layers around the Grand Staircase–Escalante National Monument, Utah, United States. *Tim Peterson*

Bottom: Tile made with native clay collected around the Grand Staircase–Escalante National Monument, Utah, United States. Shrinkage exceeded 12% and the tile warped.

A clay body is a formulation containing clays and nonclays. A simple clay body formulation, for example, can consist of equal parts of clay and sand. In a clay body formulation, the clays are selected for their ability to mature—to become dense and strong—within a specific temperature range. They may also be selected for their fired color, beige or buff or red. The nonclays are added to enhance workability, decrease shrinkage and warping, control porosity, and improve glaze fit.

It is acceptable to say clay when one really means clay body. As a ceramist, though, it's important to recognize the gap between a native clay and a formulated clay body, and to know how to bridge that gap. Note that you may occasionally encounter the term "tile body," which refers to a clay body created especially for tile.

Clay Maturity

When clay is fired until it is dense and strong enough to serve its purpose, we say it has matured. Firing turns clay into ceramic, and it can never revert to clay. It is a sobering thought that when we fire our wares, we irreversibly alter materials tens of millions of years old and in turn create products that will outlive us, either whole or as shards.

Maturation happens over a temperature range, not at a particular point. As clay is heated to maturation temperatures, it begins to sinter, which means it gets denser and its particles start bonding with each other. Sintered clay can often be surprisingly strong. As the temperature rises further, a matrix of glass crystals forms through the clay, increasing its strength exponentially. The clay is said to be vitrified. Beyond its vitrification range, the glass network disintegrates. The clay becomes brittle and bloats until—if fired high enough—it collapses into a shiny blob on the kiln shelf.

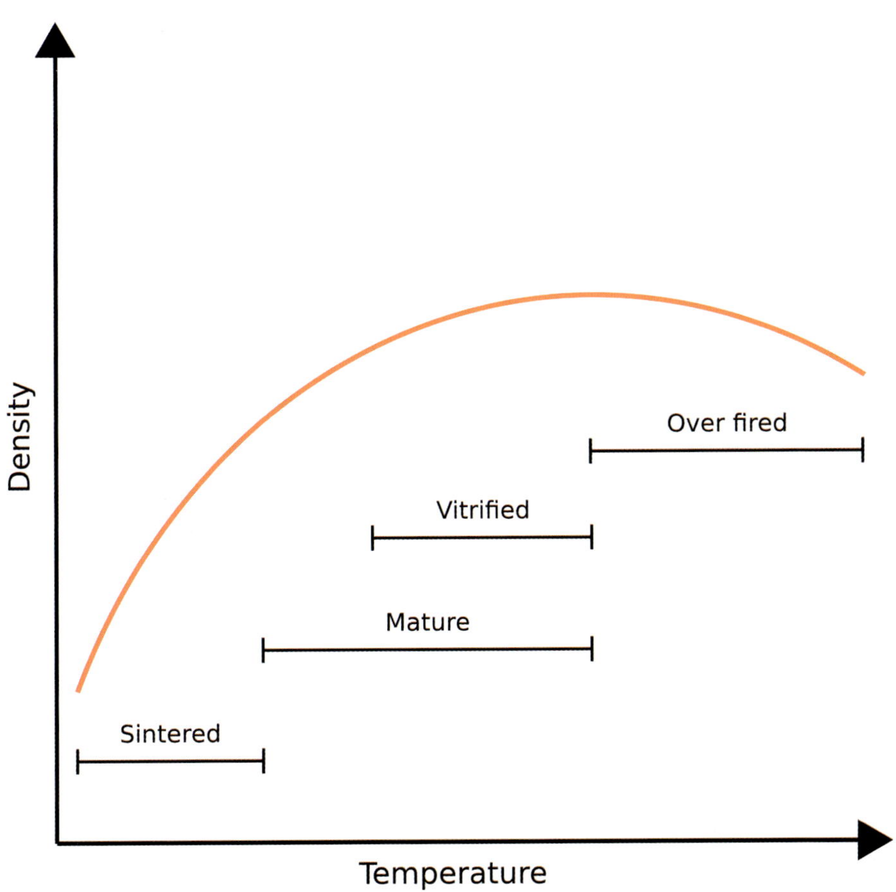

A simplified way to visualize stages in firing clay. Strength increases with density.

Above: Identical glaze work on terracotta (*left*) and white earthenware (*right*)

Detail of a mural on a terracotta body. Notice how the warm body contributes to the appeal.

What Is Terracotta?

Ceramic terminology can be confusing, and the term "terracotta" has many definitions. For our discussion here, terracotta is a grogged, red-firing body that matures in the earthenware range (950°C–1,050°C or 1,742°F–1,922°F, between Orton cones 08 and 04).

Earthenware remains porous after firing and must be glazed for watertightness. The terracotta tile body we formulate will be mature, but only partially vitrified, when fired to 1,050°C. Indeed, it may have a porosity up to 10%. Cuerda seca requires a porous substrate, so that glaze can bond securely during the glaze firing. Were the bisque tile to be completely vitrified, glaze would either not bond at all or bond so poorly that it would flake off over time.

Many of the vibrant glaze colors we associate with cuerda seca are available only in the earthenware temperature range. Therefore bright, vibrant cuerda seca requires an earthenware body, but not necessarily a terracotta body. Although one could just as easily formulate a talc-based earthenware that would fire white or beige, a warm terracotta body gives cuerda seca unmatched depth and intensity.

Formulating a Terracotta Clay Body

A terracotta tile body formulated for cuerda seca must have these properties:

1. Fire to a strong, durable, warm-toned ceramic at earthenware temperatures.

2. Be plastic, so that it can be rolled, pressed, or extruded.

3. Have green strength, also called working strength. Green strength allows tile to be handled until it is fired.

4. Dry evenly, with no discernible warping.

5. Shrink very little during firing. Under 5% firing shrinkage is ideal.

6. Remain semivitreous once it is bisque-fired. About 10% porosity is acceptable.

7. Permit a good fit between tile and glaze.

For the tile body to display all these properties, some balancing becomes necessary. For example, there should be enough grog to provide green strength and create an open structure for even drying. But too much grog makes the body short (less plastic) and therefore prone to cracks and fissures. The percentage of grog, therefore, must remain within a narrow window.

A Heart of Red

Commercial clay suppliers sometimes use pigments such as iron oxide or umber to create a red-firing clay body from lighter-firing clays. Their industrial mixers can uniformly disperse small quantities of pigments through large quantities of clay. Because such uniform mixing is difficult at studio scale, powdered pigments tend to speckle the clay body.

Additionally, native red clays contain impurities that act as fluxes. It is impossible to identify all these impurities and include them in a formulation. Therefore, a body created by adding red pigment to lighter-firing clays will demonstrate neither the firing characteristics nor the warm appeal of a true red-firing body. It is best to start with a red-firing clay at the heart of your clay body.

A starting point for a body formulation is this rule of thumb: three parts clays and one part nonclays. This ratio gives us leeway to add several nonclays to improve the working and firing properties of the clays.

Earthenware clay bodies come in a range of colors. Location: Clay Planet, San Jose, California.

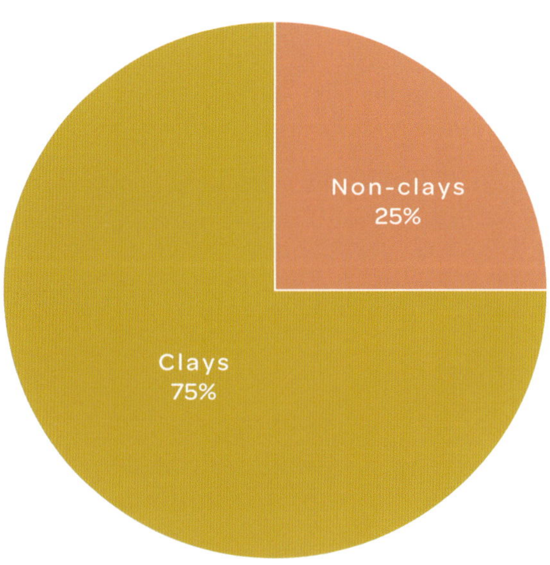

Let's start with the clays. The dominant component should be a red-firing earthenware that will yield strong, durable tile when fired to 1,050°C (cone 04). In the United States, Cedar Heights' Redart is a good choice, firing to a beautiful reddish brown. Add ball clay for plasticity and a little fireclay for tooth or roughness. Our formulation looks like this:

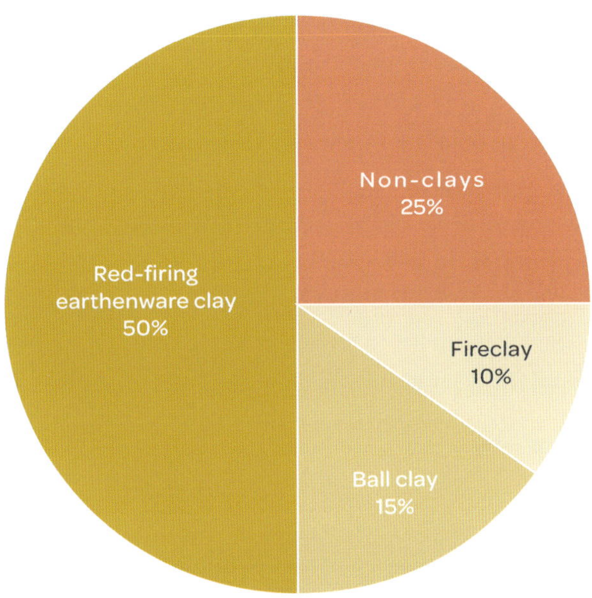

For this clay mix to yield strong and durable tile for cuerda seca, it must have good green strength and low, even shrinkage. The porosity should be kept within bounds. Glazes, including ready-made low-fire glazes, should fit well to the tile. Let's consider these requirements sequentially and identify non-clays to satisfy them.

Grog is made by grinding down high-temperature construction brick, such as this pile at Plainsman Clays, Alberta, Canada.
Tony Hansen, Digitalfire

Increase Green Strength and Reduce Shrinkage

Grog—fired clay crushed into powder—improves green strength and reduces overall shrinkage. It also creates a more open body structure, which facilitates even drying.

Grog comes in different grades—fine, medium, and coarse, identifiable by mesh number. Although mesh numbers vary by manufacturer, a larger number always indicates a finer grog. In the United States, a standard medium grog is Mulcoa 48; a fine grog, Mulcoa 200.

In India, dust sweepings from brick factories are sometimes sold as grog. This can work only if the bricks were fired to a higher temperature than the intended firing temperature of your clay body. Otherwise, the brick dust will improve green strength but will not help lower shrinkage.

We can expect 10% grog to provide the required green strength and shrinkage properties. An equal mix of medium and fine grogs creates a favorable particle-size distribution in the clay body. Our formulation now looks like this:

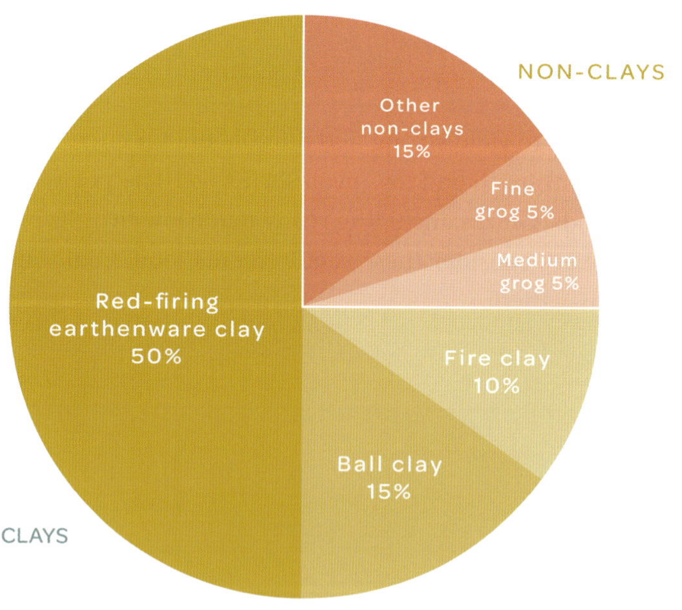

Improve Glaze Fit

As temperature rises in a glaze firing, both the clay and glaze expand, but they expand to slightly different extents. As the kiln cools after firing, they conversely contract to different extents. This difference in thermal

expansion and contraction leads to fit issues between clay and glaze. At earthenware temperatures, a common fit issue is crazing, when glaze expands and contracts to a greater extent than the clay it sits on.

Glaze fit issues can be minimized by adding talc to the formulation, since talc increases the thermal expansion of earthenware bodies. A range of 10% to 12% is a good starting point. Our formulation now reads like this:

Crazing on an antique jar contributes to its aesthetic appeal. *Monica Bariya*

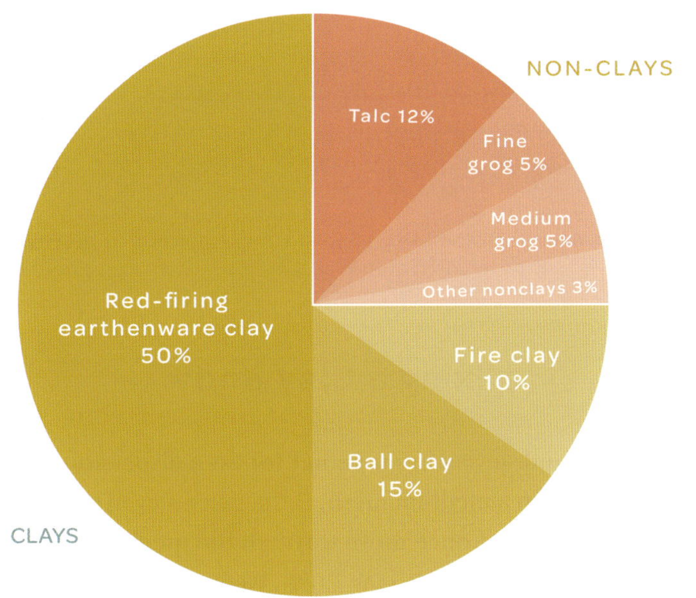

Improve Plasticity

The grog in our formulation makes the body a little short, prone to fissures and surface cracks while it is being worked. Bentonite, a very fine clay, greatly improves plasticity. We will add 3% bentonite now to complete our formulation. Note that although bentonite is a clay, it is added here to improve the working properties of the clay body.

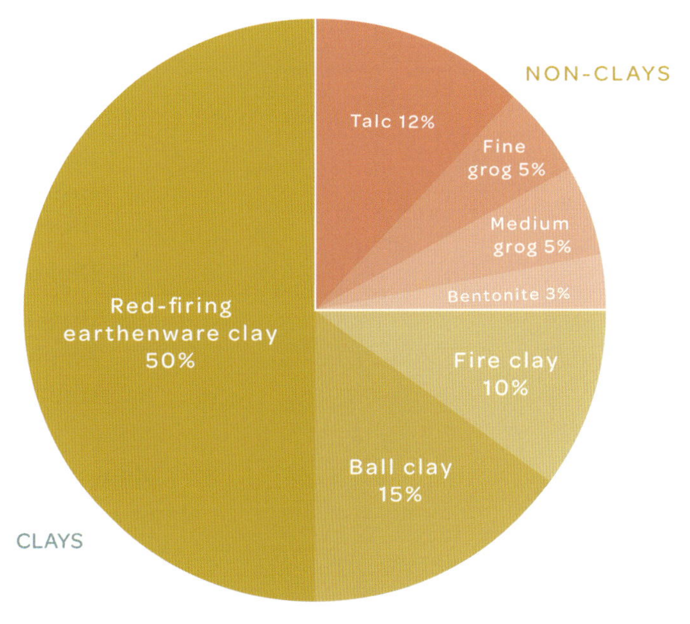

The next step is to mix up a trial amount for testing.

INGREDIENT	FOR 5 KILOS	FOR 10 POUNDS
Red-firing earthenware clay	2.5 kg	5 lbs.
Ball clay	750 g	1.5 lbs.
Fireclay	500 g	1 lb.
Fine grog	250 g	½ lb.
Medium grog	250 g	½ lb.
Talc	600 g	1 lb., 3 oz.
Bentonite	150 g	6½ oz.
Water	Approx. 300 ml/kg = 1.5 L	Approx. 5 oz./lb. = 50 oz.

Mixing a Clay Body

Weigh ingredients and sieve them together for even dispersion, first through a coarse mesh and then a medium mesh. If the twice-sieved mixture does not look homogenous, sieve it once more through the medium mesh.

Weighing and sieving raises clay dust. Wear a mask and follow these additional precautions:

1. Weigh talc out first and tip it gently into the sieve. Talc, with the finest particle size of all ingredients in this formulation, easily rises up in clouds of white dust. Trapping it under the other ingredients minimizes talc dust.

2. Do not tap the scoop against the scale or sieve.

3. If you are sieving by hand, use a gentle rotary motion, keeping your hand under the mound of clay and in firm contact with the screen.

Add water to the sieved mixture, ball it up roughly, and pack it into a plastic bag. Twist the bag closed.

Let the clay body rest for a day so that it can hydrate evenly. Then wedge it and proceed to make tiles for testing.

Top to bottom

Getting ready to weigh out clay body ingredients

Weighing out ingredients, starting with talc

Your hand should be under the mound of ingredients, working against the screen with a gentle, rotary motion.

Testing the Terracotta Tile Body

Much of the expert guidance on clay body testing is impractical at studio scale. For example, to determine whether a clay body has matured, we are advised to fire sample bars at the target cone, two or three cones below target, and two or three cones above target. By this method, five to seven firings are required simply to determine maturity. While a few studio ceramists may be motivated to do this, many end up simply not testing.

In this chapter, we focus on tests that are within practical reach of a small studio. Test tiles should be made by the same method that will be used for production.

Testing with Your Senses

These simple tests performed on bisqued tile help gauge whether the clay body has matured in the firing.

Tap the tile with your fingernail. You should hear a ringing tone. It will not be the bell-like ring of porcelain, but it should not sound dull or flat either.

Lick a dry tile. Your tongue will stick to it, but only slightly. If it sticks too much, the body is too porous and the tile is probably underfired. Compare by licking an earthenware brick, which is typically much more porous than the bisqued tile. Be sure that any surface you lick is clean and free of glaze.

Scratch the surface of a tile with your fingernail. Your nail should not leave a permanent mark.

Try to break the tile with your hands. It should not yield. If it breaks, it is underfired.

Testing for Shrinkage

Green tile shrinks as it dries and shrinks further when it is fired. Drying shrinkage depends on the plasticity and composition of the tile body. Firing shrinkage indicates maturation of the clay body. Total shrinkage, the sum of the two, determines how green tiles must be sized to compensate for shrinkage.

To calculate shrinkage, roll out three tiles. On each, mark a length of 100 mm, like this:

Measure the length in mm between the lines after the tiles are bone dry. This is the dry length.

Drying shrinkage, as a percentage = 100 – dry length.

Now fire the tiles to cone 04. Once again, measure the length in mm. This is the fired length.

Firing shrinkage, as a percentage = dry length – fired length.
Total shrinkage, as a percentage = 100 – fired length.

Average the measurements from the three tiles to get a single value for total shrinkage. Here are shrinkage calculations for the body we formulated earlier.

100 mm marked on green tile

	DRY LENGTH (mm)	FIRED LENGTH (mm)	DRYING SHRINKAGE (%)	FIRED SHRINKAGE (%)	TOTAL SHRINKAGE (%)
Tile 1	97	95	100-97=3	97-95=2	3+2 = 5
Tile 1	97	95	100-97=3	97-95=2	3+2 = 5
Tile 1	95	94	100-95=5	95-94=1	5+1 = 6

Taking the average, TOTAL SHRINKAGE (5 + 5 + 6) / 3 = 5.3%

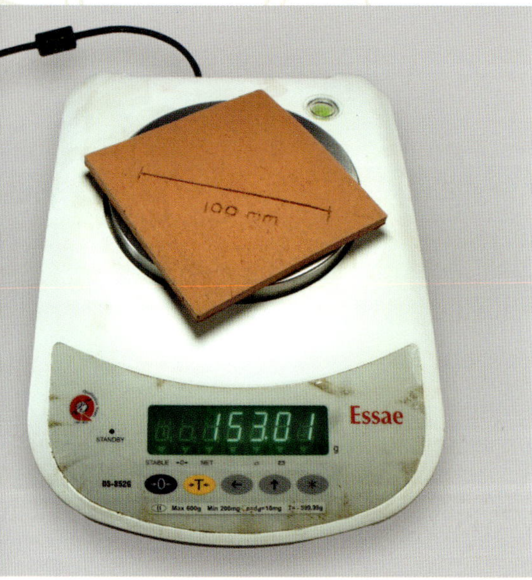

Top: Tiles in stockpot, undergoing the water absorbency test

Bottom: Weighing boiled and soaked tile for water absorbency

Testing for Water Absorbency (or Porosity)

The amount of water absorbed by a fired tile indicates its porosity. Earthenware tile will always remain a little porous after firing, but too high a porosity indicates insufficient strength.

Weigh three dry bisqued tiles individually, in grams. Record these weights as D1 (dry weight 1), D2, and D3.

Place the tiles in a large stockpot filled with water. Remember which tile is which, taking a photograph of their positions in the pot if necessary. Tiles must be completely submerged.

Simmer the tiles for an hour, topping up the water as necessary. Tiles should remain submerged the entire time. Then turn off the heat and leave the pot undisturbed for twenty-four hours.

Take each tile out of the water, surface-dry it with a cloth, and weigh it immediately. Record the weights in grams as W1, W2, and W3.

For tile 1, water absorbency or porosity percentage is [(W1-D1)/D1] x 100.

If the wet weight for tile 1 is 571 g and its dry weight was 530 g, then its porosity is [(571 − 530) / 530] × 100, which equals 7.7%.

Average the results of the three tiles to arrive at the porosity for your clay body.

Testing for Strength

Although testing for shrinkage and porosity is usually sufficient, you may occasionally need to measure tile strength. For this, tiles must be sent to a laboratory, which can run a transverse-strength test, sometimes called a modulus of rupture test. The testing instrument applies gradual, even pressure across the face of the tile and records the point at which the tile breaks into two.

Fine-Tuning the Tile Body

This chapter covers common tile body issues, with solutions for minimizing or eliminating each issue.

Small Cracks and Fissures When Tile Is Rolled Out

Small cracks or fissures may sometimes appear in green tile. This is more likely when tile is formed by rolling than when it is pressed from a mold.

Close-up view of a short clay body

Through a slab roller, clay is compressed and squeezed. In a press mold, clay is only compressed. *Mohini Bariya*

Your first response to this issue might be to increase bentonite in the formulation. There is, however, an upper limit to how much bentonite can be used. Increased much beyond the 3% in our current formulation, bentonite will slow down drying of the green tile. Too much bentonite and the tile may not dry for weeks.

Plasticity can instead be improved by adding sodium lignosulfonate, a free-flowing brown powder that is a byproduct of the wood pulp industry. Just 1% "ligno" dramatically improves plasticity. Because ligno is hygroscopic, rapidly absorbing moisture from the air, it cannot be sieved along with the dry ingredients. It must instead be dissolved in the water used to make up the clay mass. The percentage of bentonite in the formulation can be brought down to compensate for the ligno.

Ligno must be stored in a dry area, double bagged. In spite of these precautions, it can quickly turn into a sticky mass resembling amber. Fortunately, its plasticizing properties remain unchanged.

Tiles Are Too Porous

Since earthenware is by nature porous, there's a limit to how much porosity can be reduced. Nevertheless, bisque tile porosity should hover around 10% at most, to ensure that the tile is strong and durable.

Adding a little frit to an earthenware tile body aids maturation. Ferro Frit 3124, a high–calcium borosilicate frit that starts melting at around 900°C, is suitable. Replacing 3%–5% of talc in the formula with frit will bring porosity down by a few percentage points. Once porosity approaches 10%, shift your attention to identifying a good glaze that will fit well to the tile.

Breakage Is Too High

If you notice a lot of tile cracking in the bisque firing, the thermal shock resistance of the tile body may need improvement. Several minerals are available to improve the shock resistance of an earthenware body.

Wollastonite, a mineral with an acicular crystal structure, works particularly well. It also enhances green strength and aids even drying. Up to half the talc in the body formulation can be replaced by wollastonite.

Needle-shaped crystals of wollastonite photographed at Monte Somma, Campania, Italy. *Wikimedia Commons*

Bisque Tiles Have a Scummy Surface

Efflorescence or scumming, a white surface patchiness, is caused by soluble sulfates migrating to the tile surface. Adding barium carbonate to the tile body converts soluble sulfates to insoluble barium sulfate, arresting migration. Scumming needs correction only if a portion of the tile is to remain unglazed. A very small amount of barium carbonate, 0.5%, is sufficient. It should be dissolved in the water used to mix up the clay.

SECTION 3

MAKING BISQUE | 10: Making Flat Tile | 11: Firing a Bisque Load

Making Flat Tile

There are three deceptively simple requirements for making flat tile at studio scale:

1. Clay must be wedged very well.
2. The freshly formed slab should not deform—bend or twist—in any way.
3. The tile must dry evenly.

Any method that satisfies these three requirements is viable. Here we use a slab roller, but you could press or extrude tiles as well. The slab roller method is suitable for the individual ceramist or for a small studio. If you do not as yet own a slab roller, you can hand-roll slabs, using wooden slats as guides. For production, press molds on a ram press provide better throughput.

Correcting Clay Consistency

Start with firm, stiff clay. Although stiff clay is hard to roll, the resulting slab is less prone to warping than if the clay were limp. With short clay bodies, it's easy to tell when they're not stiff enough. The clay feels flabby, and tiny fissures appear while wedging. If this happens, rest the clay on a plaster bat to firm up.

Clay resting on a plaster bat for stiffening up

Setting Up

Right: Setting up to roll out slabs—weighing scale, water-resistant plywood boards slightly larger than desired tile size, trimming scalpel, clay stamp, sponge, stiff card

Below: This antique wall tile from Multan is 9" square and ¾" thick.

Weigh and Wedge

Handmade tile must be thicker than machine-made tile, so that it is strong and flat. Historically, handmade tiles have been up to an inch thick.

For tiles 4" (10 cm) square or smaller, a ⅜" (1 cm) thick slab is adequate. For larger tiles, roll slightly thicker ½" (1.3 cm) slabs.

Estimate how much clay you need per slab. The slab should be large enough to comfortably yield a tile, but without too much surplus to rewedge. For 4" square tile, you need approximately 600 g (1.3 lbs.) per slab; for 6" tile, 1.2 kg (2.6 lbs.).

Roll Out Slab

Fold the slab roller canvas in half, tuck the folded edge between the rollers, and flip it open, ready to receive clay.

On your workbench, roughly flatten a ball of clay into an ovoid, using a mallet or the side of your hand. Bring your strokes down vertically on the face of the clay. Angled strokes will stretch the clay unevenly, increasing the chances of warped or deformed tile.

Above: Weighing the estimated amount of clay per slab

Left: Hand (or mallet) must hit the clay down at right angles. *Mohini Bariya*

The length of the ovoid should approximate a side of the tile. It is not advisable to work directly on the slab roller since the rollers may get misaligned over time.

Cradle the clay between your hands and transfer it to the slab roller.

Beat down the leading edge so that the rollers engage the slab easily. Cover the clay with the canvas and roll the entire assembly through with a smooth, even movement. Then roll it back in the opposite direction.

Peel back the top layer of canvas to reveal the slab. The surface you see will be the back of the tile, which you may backstamp now. Do not incise grooves on the back just yet. Grooves will be incised later, when the tile is leather hard.

1. Canvas ready on slab roller

2. Flattened ball of clay. The tile cutter at rear acts as a gauge for sizing the ovoid.

3. Ovoid of clay on slab roller canvas. Notice the thinner leading edge.

Right: Roll the clay through in both directions for a more uniform slab. *Mohini Bariya*

Make it a practice to observe the thickness of the freshly rolled slab. If the roller adjustment is uneven, the slab will look like this:

Trim and Flip

Place a piece of plywood board on the slab. Trim away any excess clay around the edges of the board. If trimmings are too large, consider wedging slightly smaller balls of clay next time around.

Firmly pull the sides of the canvas over the board, leaving no slack between canvas and board.

A tight grip on the plywood-slab-canvas package prevents deformation as the slab is flipped over onto the board. Holding the package firmly, flip it over. Peel away the canvas carefully to expose the slab.

Smooth and Finish

Back at your workbench, smooth out the slab face, first with a damp sponge and then with a ribbed or stiff plastic card.

Smoothing compresses the clay and removes texture marks caused by the canvas. Errant air pockets also reveal themselves as small bumps on an otherwise smooth surface. Gouge air pockets with a scalpel or needle tool, repair with tiny wads of clay, and then smooth over again.

Smoothing is not required if you use SlabMat instead of canvas. However, you still need to examine the slab for air pockets. Let the slab rest for about ten minutes before cutting out tile.

4. Trim away excess clay.

5. Fold the canvas firmly over all four sides of the board, leaving no slack.

6. Smooth over the surface.

7. Eliminate air pockets.

8. Hold your hand against the edge to release tile gently onto the board.

9. Telltale lifting of tile corners. Left unattended, these tiles will warp.

Cut Tile and Dry to "Leather Hard"

Cut tile out with a tile cutter and gently release it onto a board. Hold your hand against the edge as you release the cutter, so that the tile does not lift along with the cutter.

If tile does not release easily, let slabs rest a little longer before cutting. Sprinkling talc on the cutter face also eases release.

The freshly cut tiles should remain undisturbed on their boards for a few hours. Watch the corners. If they start lifting, they may be drying too fast. Cover loosely with a plastic sheet to slow down drying.

To check if tiles have dried to leather hard, place a board on top of a tile and invert the board-tile-board sandwich. If the tile releases, you can move on to the next step. If it does not, wait a few more hours. Once you've done this a few times, you will develop a sense for how quickly tiles release from their boards under your studio conditions.

Score, Finish, and Grid-Dry

Carefully flip the leather-hard tiles over and score the backs. Scoring increases the surface area for adhesion.

Smooth the tile edges and place tiles on a grid rack until they are bone dry. It is not necessary to keep flipping the tiles or moving them around, because the grid rack ensures even air circulation.

It can take several days for the tiles to get bone dry, ready for firing. If a tile pressed against your cheek feels cool, it is not yet bone dry.

Augmenting a Slab

Sometimes a slab rolls out just a little too small to yield a tile. This is more likely when you're making larger tiles. Instead of rewedging and rerolling the slab, you can simply augment it to the required size. To do this, mark a faint outline of the tile cutter on the slab, to know where you need to add clay.

Trim a small piece from an area of surplus. Score and slip the two edges to be joined.

Smooth the join. Adjusting the slab roller down by the tiniest adjustment possible, roll the slab through again. This makes the tile slightly, but imperceptibly, thinner. The slab is now large enough to yield a tile.

10. Backs of green tiles, scored when tiles became leather hard

Above right: A repurposed office letter rack provides even circulation for drying green tiles.

11.

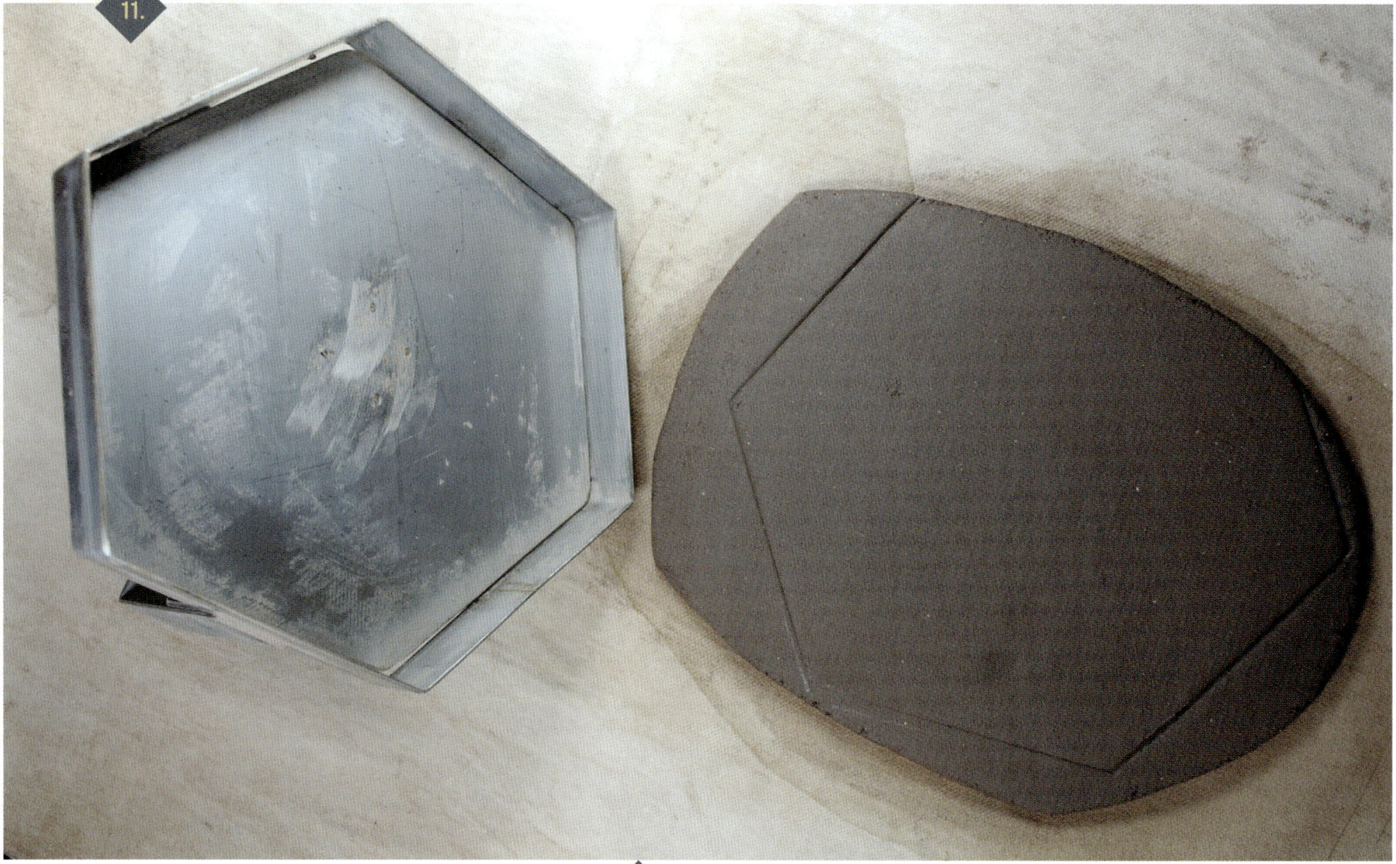

12.

11. Outline of tile cutter shows where more clay is required.

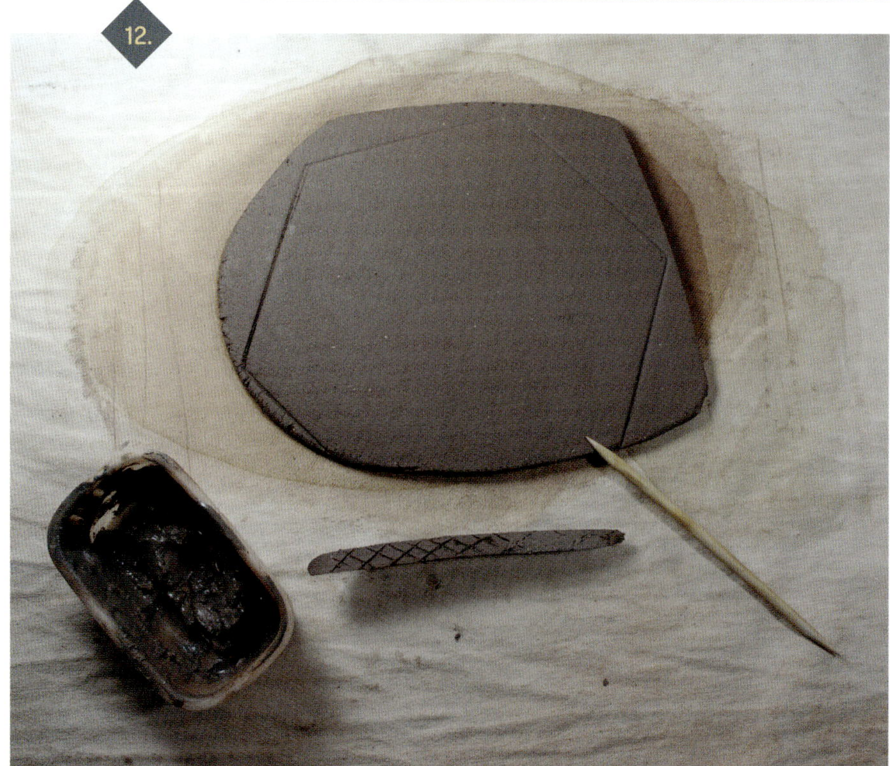

12. Slip and score both surfaces.

Firing a Bisque Load

The slowest step in the cuerda seca process, and the one most likely to test your patience, is drying green tile. Drying times vary wildly depending on the season and your studio conditions.

Confirm that tiles are bone dry by holding one to your cheek. It should not feel cool to the touch. An unfired bone-dry tile from an earlier batch can serve as a reference. Terracotta clay bodies often show a distinct color shift between leather-hard and bone-dry stages, and this too can be used to gauge dryness.

The Indian monsoons are beautiful, but tiles dry very slowly in such weather.

Reclaim any tiles that are even slightly warped. Since small warps will only get exacerbated in firing, resist the temptation to fire imperfect tile.

Loading the Kiln

In a small kiln, green tiles can be stacked in pairs, face to face. Sprinkle sand on the shelves before setting down the tile. Avoid stacking any higher, since the lowermost tiles may crack.

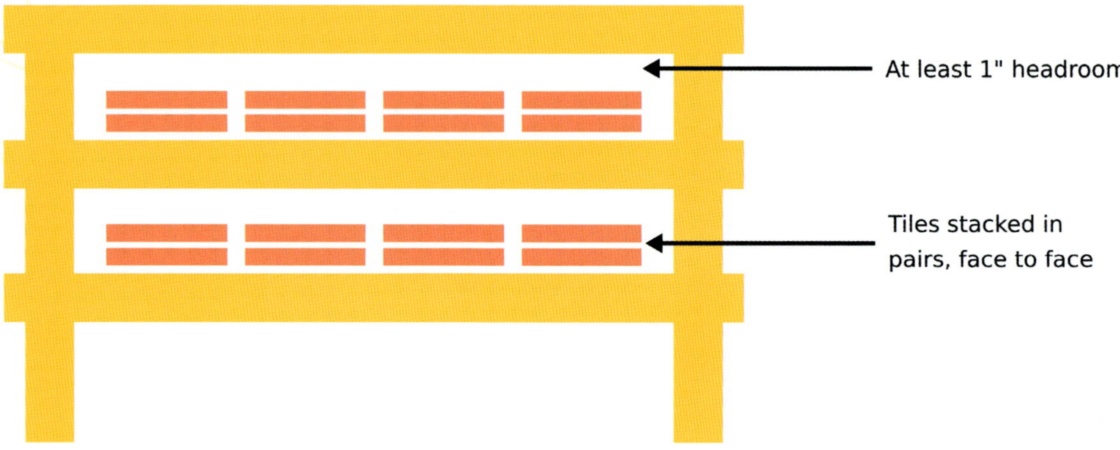

In a larger kiln, tiles can be supported bookshelf style on squat posts.

Top: Tiles loaded flat in the kiln, each pair face to face

Bottom: Tiles supported by posts for bisque firing

Allow tiles an inch of headroom, and leave some space around the thermocouple. Kiln furniture designed for tile (rods, stands, or setters) makes loading and firing more efficient.

The Firing Cycle

Bisque firing is slow and steady. At the early stages of firing, there must be an escape route for residual moisture. If a piece of glass (or eyeglasses, if you happen to wear them) held by the peephole fogs up, moisture is still venting. It's good practice to leave the peephole open until all the moisture has vented.

Make it a practice to place cones in the kiln. Cones are the best indicators that your firing cycle is on track. Tabulated here is the bisque-firing cycle for my Skutt KM1027, in °C and °F. The slight variation between the two is simply to make it easy to program the kiln controller. It will not affect the results of the firing.

FIRING CYCLE (TEMPERATURES IN °C)

RAMP AND SOAK	TARGET TEMPERATURE	TIME	NOTES
Ramp at 50°C/hour	150°C	About 2.5 hours, depending on the ambient temperature when firing began.	Lid is propped open for the first two hours. Top peephole is open.
Soak	Steady at 150°C	10 minutes	Brings the entire load up to this temperature.
Ramp at 150°/hour	975°C	5.5 hours	At 300°C, if no moisture is escaping, close the top peephole.
Soak	Steady at 975°C	15 minutes	Brings the entire load up to this temperature.
Ramp at 150°/hour	1050°C	0.5 hour	
Final soak	Steady at 1,050°C	0.5 hour, or until cone indicates end of cycle.	

AN EQUIVALENT FIRING CYCLE IN °F

RAMP AND SOAK	TARGET TEMPERATURE	TIME	NOTES
Ramp at 80°F/hour	300°F	About three hours, depending on the ambient temperature when firing began.	Lid is propped open for the first two hours. Top peephole is open.
Soak	Steady at 300°F	10 minutes	Brings the entire load up to this temperature.
Ramp at 300°F/hour	1,800°F	5 hours	At 550°F, if no moisture is escaping, close the top peephole.
Soak	Steady at 1,800°F	10 minutes	Brings the entire load up to this temperature.
Ramp at 300°F/hour	1,970°F	35 minutes	
Final soak	Steady at 1,970°F	0.5 hour, or until cone indicates end of cycle.	

Above: Tile is darker at the center.

Right: Marbling becomes apparent after firing.

Troubleshooting

Here are some issues that may crop up after bisque firing.

Tile Shows Dark Center

This is the result of a temperature gradient when tiles are fired in pairs, face to face. Usually this does not matter, because the surface will be covered by glaze. For even surface color, stack tiles singly or vertically on shelves or use kiln rods and setters.

Bisque Tile Looks Marbled

This is due to inadequate mixing and poor wedging of the wet clay. It may not be readily discernible in the green tile.

Visible Cracks in Bisque Tile

In square or rectangular tile, visible cracks in bisque may appear due to several reasons.

1. The clay body has insufficient resistance to thermal shock.
2. Unsteady firing, usually ramping too fast.
3. Rapid cooling after cycle is completed. Typically, a midsized load should require eighteen to twenty-four hours to cool down. Check kiln insulation and make sure the kiln lid or door seals securely.
4. Stress induced while loading. To correct this, stack fewer tiles on each shelf.

Microcracks in Bisque Tile

To check for microcracks. submerge a bisque tile in water. Bubbles rising along a straight line indicate a microcrack. Bisque with microcracks must be discarded, since cracks will worsen if the tile is glazed and fired.

Above: Smaller stacks may resolve the cracking issue.

Right: A line of bubbles indicates a microcrack. *Mohini Bariya*

SECTION 4

Design Considerations and Constraints

Inspiration for cuerda seca can come from many sources. Bas-relief on old buildings, patterns from stained glass, and motifs woven into folk textiles translate well to cuerda seca. Antique tiles—cuenca, encaustic, and majolica—are a treasure trove of design as well. Whatever its source, any design selected for cuerda seca must have bold lines and distinct areas that can be represented in flat colors. Designs with cross-hatching, color gradients, or subtleties of light and shade are not suitable. A look at some examples will clarify these requirements.

Here is a simple example. This antique tile was painted with underglaze pigments and then dipped into a clear glaze. The watercolor-like underglaze is particularly noticeable in the turquoise flowers. Nevertheless, because the lines are well defined, the pattern can translate to cuerda seca.

The painterly effect of underglazes is now replaced by bold cuerda seca, with the pattern lending itself well to both techniques.

Above left: Antique underglaze tile from Multan

Above right: A digital rendering of the pattern, preparatory for cuerda seca tile

The Multan pattern in cuerda seca

From Vintage Poster Art to Cuerda Seca Tile

Here is a more complex translation. The central image in this evocative poster shows high contrast and clearly demarcated areas of flat color. There is very little shading (notice the gray on the woman's blouse), and there are virtually no gradients. The colors can be closely matched to earthenware glazes. It appears to be a good candidate for cuerda seca.

The first step with an image like this is to overlay a grid simulating tiles and grout lines. Visual thinkers may do this mentally; the rest of us must use a computer or a paper printout. Grid dimensions are somewhat arbitrary at this stage, on the basis of a judgment of how much of the image can be accommodated on each tile.

The grid should be positioned so that details in the picture do not fall on grid lines. Sometimes, minor details have to be eliminated, although this does not appear necessary here.

Above left: Vintage screen-printed travel poster. *Wiki Commons*

Above right: Grid overlay to represent tiles and grout lines

The grid overlay also helps compose the mural. The mural could, for example, be composed in a landscape format, wider than tall. In a landscape composition, the river, village, and mountains add to the sense of tropical ease.

Alternately, the mural could be composed in portrait format, taller than wide. A portrait format focuses on the woman rather than her surroundings.

After the composition is finalized, relevant parts of the picture must be scaled, digitally or using a photocopier, to fit the available bisque tiles.

Important details, such as the woman's eyes, for example, must not fall into the cracks.

A composition in landscape format

A composition in portrait format

Scale

Every pattern has a size range within which it is most appealing. Before committing a design to tile, print it out at the intended size and examine the printed design. Is it visually pleasing at this size? Is there enough complexity to engage the viewer or—conversely—is it too cluttered? Such judgments are better made using paper printouts, rather than computer displays.

Boosting Visual Interest

If large areas of a tile or mural are glazed in a single color, visual impact and appeal are reduced. Especially in large murals, it helps to subdivide large areas and glaze them in colors that complement each other. A good example is the floor-to-ceiling post office in Monterey, California. Clouds, the sea, trees, and grass all have been glazed in multiple colors to enhance impact. This dual-color treatment succeeds on tile that will be viewed from a distance, because the viewer's eye melds adjacent color areas into a coherent whole. It does not work well on tile that is viewed up close.

Above left: This relatively simple pattern works well on small tiles, up to 4" square.

Above right: This intricate pattern is beautiful on tiles 6" square and larger.

Mural in the post office building in Monterey, California. A public artwork commission under the New Deal Program, 1934–1943. Artist: Stanton Willard. *Venkatesa Chandrasekaran*

From Inspiration to Pattern

In a cuerda seca pattern, all lines must form closed areas, with the exception of lines that touch tile edges.

Inlay border in the Taj Mahal made by the *parchin kari* technique. Semiprecious stones such as agate, onyx, jasper, and jade are sliced, laid in marble, and polished to perfection.

In this chapter, we will convert a design into a line drawing for cuerda seca. Our inspiration is an inlay border from the Taj Mahal.

The first step is to take a high-resolution, eye-level, close-up photograph of the design. This is straightforward when the design comes from a book or piece of fabric, but not when it is high up on a building. Distorted photos must first be corrected using photo-editing software.

If the pattern repeats or tessellates, identify its core element.

To reproduce the pattern manually, resize the core element on a photocopier and transfer it to tile using the old-fashioned standby, carbon paper. The cuerda seca lines will have to be painted individually on each tile. To make the task less onerous, you could consider screen-printing the cuerda seca lines (screen printing is discussed in section 5) and converting the pattern to a vector—a digital line drawing—on the computer.

Creating a Vector Pattern

Reproducing a pattern in vector format offers many advantages. The pattern can be scaled up or down to fit several sizes of tile, without distortion or loss of detail. This makes a vector different from a pixel-based image, which gets fuzzy as it is enlarged. Inkscape, an excellent drawing program, is available as a free download from www.inkscape.org.

Although a vector-drawing tutorial is outside the scope of this book, here are a few tips using Inkscape.

Use the Layers feature and import the photo into the base layer. Bring down the opacity of the photo to about 50% to make outlines clearly visible.

Reduce the photo layer opacity so that outlines become easier to trace.

Add a second layer with guide and grid lines for reference. Lock the photo and guide layers. In a third layer, trace outlines with the pen tool. Zoom in as much as you need to for accurate tracing.

Zoom in to capture detail.

The complete vector looks like this.

Once you have a vector file, several possibilities open up. For example, the stroke width can be changed to suit the size of the tile and the effect desired. A thinner stroke lets glazes dominate. A thicker stroke makes lines prominent.

Change stroke width as desired.

When the Taj Mahal inlay border is fitted to a square tile, there is open space at the top and the bottom. To fill the tile more, the aspect ratio of the pattern can easily be altered.

Fit to square.

ORIGINAL

FIT-TO-SQUARE

It's also possible to split lines into two. In cuerda seca, split lines alter the appearance of the tile significantly, since a thin line of glaze can now be trailed between each pair of lines.

Finally, with a vector file, you can simulate a tessellation and test out glaze colors, like this, before committing to clay.

Above left: Inkscape's stroke-to-path feature allows lines to be split, for glaze to be trailed in between.

Above right: Testing colors

Composition

Rendering any scene in cuerda seca, whether directly from nature or from a photo, requires tightening, simplifying, and slightly dramatizing the key elements of the composition so that they marry cuerda seca's vibrant aesthetic. While writing this book, I worked on a mural of the Oval Maidan, a cheerful and well-used public space in Mumbai, India. Many of the composition tips that follow use the Oval Maidan mural as an example.

Orientation

Mumbai's Oval Maidan, much loved by its citizens. *Suhita Shirodkar*

A cuerda seca mural that hopes to capture the spirit of the Oval Maidan must include two key elements: the Victorian Gothic skyline with its iconic Rajabai Tower, and the omnipresent cricket games on the lawns.

The orientation of the mural—whether landscape or portrait—depends on the mood to be conveyed. To create a sense of leisure and open space, the mural should be oriented as landscape, wider than tall. For a more dramatic composition emphasizing the skyward sweep of the Rajabai Tower, the mural should be oriented as a portrait, taller than wide.

Engaging the Viewer

Always try to draw your viewer into your mural. In the Oval Maidan mural, for example, the cricket game was arranged so that the players in the foreground have their backs to the viewer. This creates a line of sight that pulls the viewer's gaze into the game and then up the tower.

Sizing the Mural

The size of a mural is most obviously influenced by the wall on which it will be installed. In the case of the Oval Maidan mural, however, it was the size of the kiln. The mural had to be fired in a single firing to maximize color consistency. On the basis of the capacity of my kiln, I decided on a final size 2.5 feet wide and 3 feet tall, each tile being 6" square. The border tiles were fired separately.

Once the size was finalized, smaller figures had to be repositioned so that they did not span tiles in both directions.

A digital rendering of the final composition is very useful. Although not strictly necessary, a color-complete rendering is a handy guide.

A strong line of sight energizes a composition.

Above: As much as possible, small figures should not span grout lines.

Right: A colored guide like this is useful while screen printing and glazing.

Color

Optimizing the Glaze Palette

Although there is a breadth of glaze color available at earthenware temperatures, it is still limited compared to watercolors or pastels. You may find that cool colors—blues and greens—are more consistent and reliable than oranges or reds. A mural design should be selected that keeps the glaze palette in mind.

If high contrast or visual impact is desired, adjacent colors should be distinct in value (lightness or darkness) as well as color. This is particularly important when a mural is to be viewed from a distance. In the Oval Maidan mural, the central band of foliage was divided into areas of different colors and values so that visual interest remained strong even from a distance.

If a project has naturalistic elements, those elements should be rendered in color as close to natural as possible. For example, the kingfisher in the mural here would be unrecognizable if its glaze colors did not mimic its natural plumage.

Sometimes it becomes necessary to take creative license. In the kingfisher mural, with two blue glazes earmarked for the bird and a third for the water, there were not enough unique blues available for the sky. The solution was to use warm glazes for a kaleidoscopic horizon and sky.

Above: Varying color and value to maximize visual interest

Left: Naturalistic color or creative license

Above left: The shadows are made with resist rather than black glaze.

Above right: Using near black instead of true black, as in the use of purple for the beak, keeps resist lines well demarcated.

Opposite page

The Oval Maidan mural, ready to be installed

Using Black

When a cuerda seca mural has areas that must be rendered in black, there is some risk of losing definition where the black glaze meets resist lines. We have a few ways to address this.

Consider using resist for the black areas rather than glaze. In this example, the matte resist renders the players' shadows better than black glaze.

A second option is to use near-black glazes rather than jet black. The grape glaze used for this kingfisher's beak is less than realistic but creates the impression of a dark beak without visually blurring into the resist lines.

You could also consider using iron oxide rather than manganese dioxide for the resist paste, since this will result in slightly brownish-black lines.

Glaze Color Schemes

Both the kingfisher and Oval Maidan murals used Duncan Envision cone 06 glazes. Glaze colors are listed here.

Kingfisher

BODY: 1048 Delphinium with highlights of
1056 Agua Fresca | 1100 White | 1053 Harvest |
1204 Orange

BEAK: 1055 Grape

FEET: 1074 Cranberry

GRASS: 1205 Neon Green

TREES: 1058 Clover | 1081 Brown

RIVER: 1079 Turquoise with 1056 Agua Fresca near
horizon line

HILLS: 1060 Bright Wine

SKY: 1074 Cranberry | 1072 Baroque Gold | 1053
Harvest (without intervening resist lines)

Oval Maidan Mural

PLAYERS

CLOTHES: 1100 White | 1033 Driftwood

SKIN: 1081 Cocoa

CAP: 1048 Delphinium

BAT: 1072 Baroque Gold

LAWNS

PITCH: 1044 Sand Diego

STUMPS: 1072 Baroque Gold

GRASS: 1078 Key Lime | 1062 Light Kiwi

FOLIAGE

PALM TREES: 1061 Light Kelp | 1081 Cocoa

OTHER TREES: 1062 Light Kiwi | 1058 Clover | 1035 Raintree Green

ARCHITECTURE

RAJABAI TOWER: 1072 Baroque Gold 1044 Sand Diego | 1083 Honeysuckle

HIGH COURT BUILDINGS: 1060 Bright Wine | 1083 Honeysuckle | 1044 Sand Diego

SKY

BLUE: 1056 Agua Fresca

CLOUDS: 1100 White | 1033 Driftwood | Orchid 1012

SECTION 5

Formulation and Techniques

The knowledge that oily substances repel water has been used across craft traditions to create designs of extraordinary beauty. Indonesian batik, for example, employs wax to trace intricate patterns that resist aqueous dyes.

Indonesian artisan traces wax patterns for batik. *Wiki Commons*

Opposite page

Photo: Julie Mott Munger

Cuerda seca is based on the same principle. A resist paste, made by mixing oil and a dark oxide, is used to create a line drawing on bisque tile. Once the lines dry, water-based glazes are applied between them. The oily lines act as barriers, preventing one glaze from running into another. In the kiln, oil burns off and oxide fuses to the tile as a rich, dark outline.

The oxide most commonly used is manganese dioxide, and the oil is linseed oil, which dries when exposed to air. Although linseed oil is not the only drying oil available, it is ideal for cuerda seca because it dries quickly—but not too quickly—and is inexpensive.

Paint or Print?

Cuerda seca lines can be either hand-painted or screen-printed onto bisque tile. Painting is not easy because the oily resist dries on the paintbrush and must continually be wiped off. Also, hand-brushed lines are inherently uneven. If you plan to paint, keep several brushes of the same size at hand, along with a pot of turpentine or brush cleaner. Brushes will have to be swapped frequently and cleaned between uses.

Screen printing produces even, elegant, controlled lines and is the method of choice for creating multiples.

Preparing the Paste

Cuerda seca paste is a suspension of manganese dioxide and a borate in linseed oil. The borate acts as a flux, fusing the manganese dioxide to the tile in the firing. Laguna borate, colemanite, or a borate frit may be used. In lieu of manganese dioxide, you can experiment with dark-firing oxides such as ferric and cobalt oxide.

Exposure to high levels of manganese dioxide can damage lungs. Although it is very unlikely that preparing cuerda seca paste will harm you in any way, wear a mask while mixing. Manganese stains fingers, but fortunately, it is not absorbed significantly through the skin.

A little cuerda seca paste goes a long way, and, because it air-dries, it must be mixed in small quantities even when the requirement is large. Experience will teach you to mix the right amount for any project. Mix when you are ready to use. The recipe here should suffice for a small project, about 2 feet square.

Opposite page, top to bottom

Either raw or boiled linseed oil works, although drying times may differ slightly.

Excess linseed oil can creep, resulting in a bare band between black line and glaze. *Mohini Bariya*

Fine lines screen-printed onto a 4" bisque tile

1. Weigh out 100 g of manganese dioxide and 10 g of borate.

2. Mix the two powders and sieve them through a fine sieve.

3. Weigh out 15 g of the powder mix into a disposable plastic bowl.

4. Add linseed oil drop by drop, stirring constantly. Slow addition of oil is important since it is easy to add too much. Aim for a smooth, glossy paste with the consistency of thick honey or molasses. You will need approximately 3 g of linseed oil, but let your senses judge.

For brush painting, it is tempting to make the paste thinner, for smoother brushstrokes. Don't do it. Linseed oil will creep beyond the cuerda seca line before it dries, pushing glaze back to create an unsightly "no-man's land" between the line and the glaze.

Line Thickness

The question "How thin can a cuerda seca line be?" is not the same as "How thin should a cuerda seca line be?" The first depends on process and skill, while the second is an aesthetic decision.

When cuerda seca resist is painted on, it is difficult to control the lines, because stroke thickness depends on the painter's skill. With screen printing, lines can be controlled. The more detailed the pattern, the thinner the line can be. The only caveat is that glaze will jump lines that are too delicate.

In the 4" square tile shown here, the cuerda seca lines are 0.27 mm thick. This is about as thin as you can go without risking glaze jump.

Since cuerda seca lines are graphic elements contributing to a tile's aesthetic, line width is not a constant. It varies with the size of each tile, the overall size of the project, the distance from which the tile will be viewed, the intricacy of the pattern, and the artist's judgment.

A good way to determine line width for any pattern is by printing it out on paper at several widths and then making a design decision, as in the following example. The four-tile marlin mural was designed in the 1920s by the Catalina Clay Products Company.

Tile size: 6" square; stroke width: 0.1 mm
Lines are too delicate. Glaze is likely to jump the lines, and the design looks weak and indecisive.

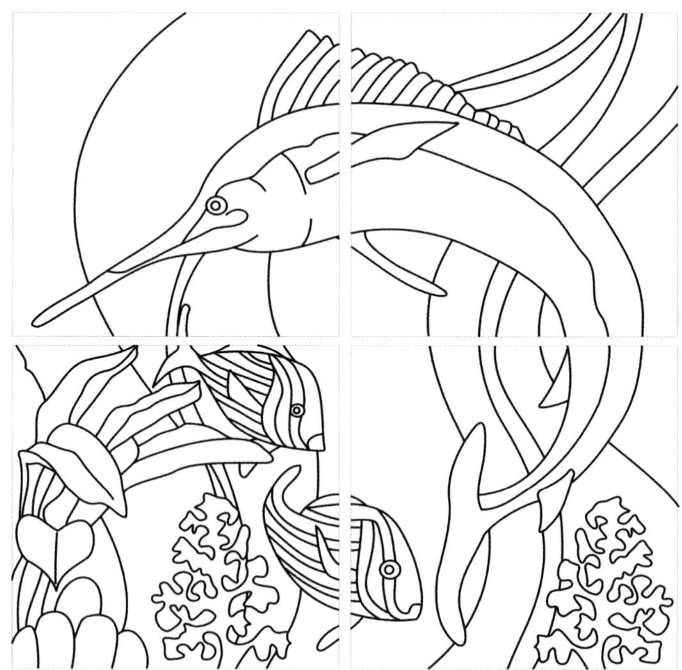

Tile size: 6" square; stroke width: 0.6 mm
Lines dominate the pattern and will overwhelm glazes once they are applied.

Tile size: 6" square; stroke width: 0.3 mm
Lines are firm and functional but do not dominate the design.

Textural Lines

By now, it should be quite clear that cuerda seca lines are closed lines that keep glazes apart. This is true for the most part, but there are some exceptions. Open cuerda seca lines can be used to add texture and detail. In this detail from the Monterey Post Office mural, textural lines create visual transitions between the light and dark blues of the bay.

Wherever cuerda seca resist is used—to separate glazes, to create texture and shadow, or simply to boost visual interest—the tile remains unglazed. It is therefore important to seal tiles before installation because adhesive or grout will be difficult, if not impossible, to rub off the unglazed areas.

Above left: A reproduction of Catalina Clay's vintage mural enlivens a doorway in Santa Cruz, California. *Julie Mott Munger*

Above right: Detail from mural in the post office building in Monterey, California. A public artwork commission under the New Deal Program, 1934–1943. Artist: Stanton Willard. *Mallika Bariya*

Screen-Printing Cuerda Seca Lines

Screen printing is a simple, elegant method for applying a cuerda seca line pattern on bisque tile. A negative of the pattern must first be imprinted on a screen, which consists of nylon or polyester mesh stretched on a wood or metal frame.

Resist is pressed through the prepared screen. It passes through the unblocked areas, which correspond to the pattern, and on to the tile below.

A screen can be imprinted manually, by applying screen blocker on the nonpattern areas. This is a laborious process, and it is difficult to achieve clean lines. The UV exposure method, which uses photosensitive emulsion, produces clean, consistent results.

Exposing a Screen

Screen exposure, also called burning, can be outsourced. Online screen shops accept artwork in several formats and burn screens in the required screen and mesh sizes. In India, neighborhood print shops will often, for a small fee, expose a screen for you. For cuerda seca, the screen should be a couple of inches larger than the tile all around. Mesh size can be between 80 and 160.

For better control and efficiency in your studio process, you may wish to expose screens in-house.

Above: The pattern to be imprinted on screen

Below: The negative pattern imprinted on a screen

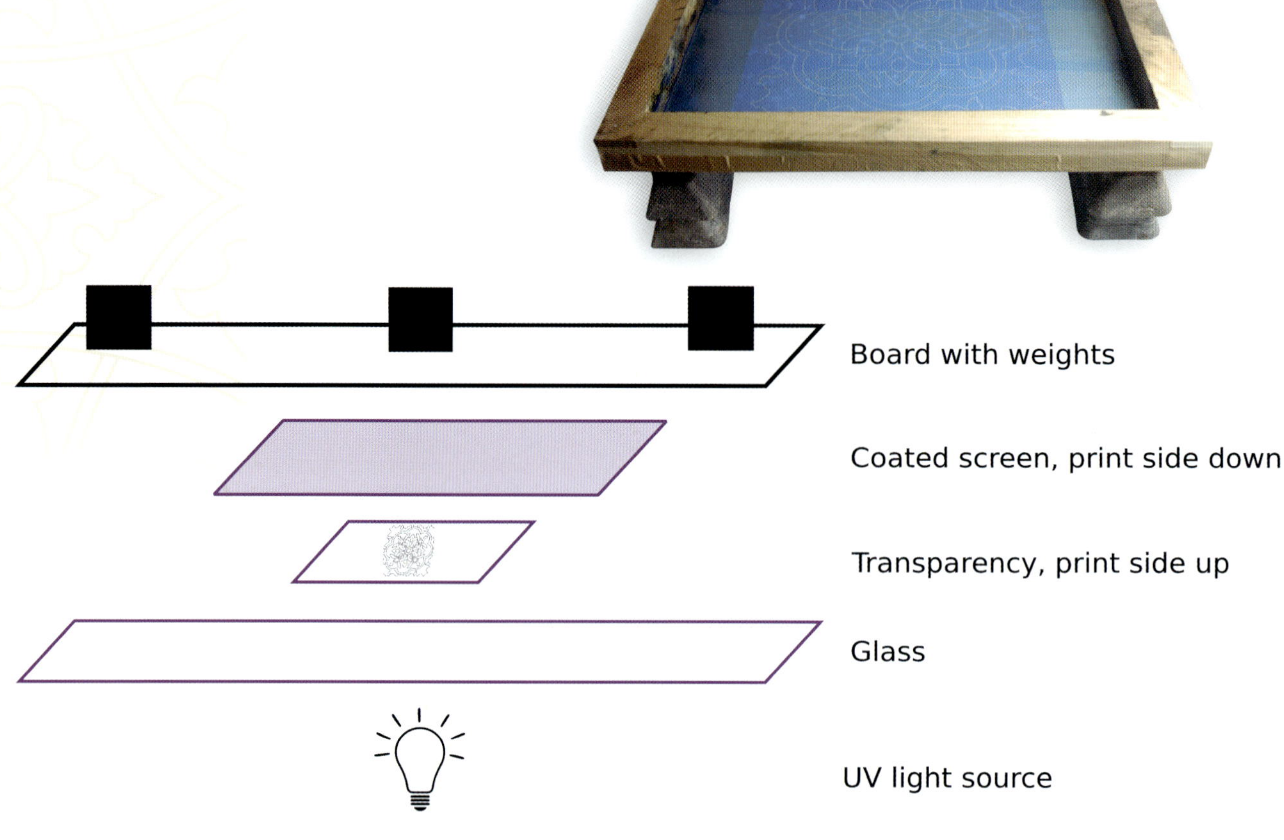

Although this is not complicated, it is difficult to provide precise instructions here because there are many variables, and because exposure times depend on the setup and emulsions used. The basic principles of screen exposure are outlined here. For details, please refer to a standard screen-printing textbook.

Screens must be exposed in a dark room, with a UV exposure table. There should be a faucet close by to wash exposed screens.

The screen is coated with photosensitive emulsion, first the print side and then the ink side, and allowed to dry. The pattern, printed on a transparency, is placed on a UV exposure table, and the dry screen is centered on top of it, print side down. The entire assembly is weighed down to maximize contact between pattern and emulsion.

UV light is then turned on for a predetermined amount of time. This may range from a few seconds to several minutes, depending on the setup. The photosensitive emulsion hardens in the exposed areas but remains soft in the unexposed areas, which were masked by the pattern. The screen is then washed in a strong jet of water.

Top right: In this picture, the ink side of the screen is on top, and the print side is below.

Top left: Exposing a screen

Soft emulsion washes away, leaving a negative of the pattern imprinted on the screen.

Factors affecting exposure time include the intensity of the UV source, the distance between UV source and screen, the type of emulsion and how old it is, and the mesh size of the screen. If exposure time is inadequate, the screen is underexposed, and all the emulsion gets washed away in the jet of water. If it is overexposed, even those parts that should have remained soft (masked by pattern) become hard, and so nothing washes away. Either way, the screen must be scrubbed and redone. Within a few iterations, you will arrive at the correct exposure time for your setup and emulsion.

Rigging a Table for Screen Printing

Any table or desk can function as a screen-printing table, by hinging a strip of wood along its short edge. Hinges should be firmly screwed into the table, since even the slightest give will cause registration errors. The table should be well lit since shadows make it difficult to pull perfectly registered prints.

Setting Up

You might wish to wear disposable gloves while printing, to protect your fingers from getting stained. For multi-screen murals, keep a reference image at hand.

Above: Rigging a desk or table for screen printing. *Mohini Bariya*

Left: Resist, screen, C clamps, squeegee, cleanup solvent and rags, justification L, test print tiles. The cuerda seca resist should be mixed after you've set up your screen, just when you're ready to print.

Making a Registration Guide

An L square fixed to the printing table helps register each tile before printing. Any two straight edges can be taped at right angles to form an L. The L should be thinner than the tile so that it does not impede contact between screen and tile.

Readying Test Tiles

You need not sacrifice bisque to pull test prints. Prepare a few test tiles by taping white paper around bisque tiles.

Top left: A simple L square made by taping two pieces of broken tile at right angles

Top right: Tiles with white paper firmly taped to them for pulling test prints

Bottom: Screen fixed to table with C clamps

Printing

Fix the screen firmly to the table with C clamps, making sure there is no play or wobble. Angle the clamps so that they do not get in the way of the squeegee.

Place a test tile on the table and align it with the pattern on the screen. You may need to nudge the tile with a chopstick or ruler (since your fingers

won't reach into the narrow gap) until it is perfectly aligned.

When the tile appears perfectly aligned, lift the screen and, holding the tile in place, nudge the L against one of the lower corners of the tile. Tape the L to the table.

Lower the screen down gently on the tile. You are now ready to pull a test print.

Trail a line of resist along the far side of the screen. Resist should not extend much beyond tile edges, since excess resist on the screen is wasteful and will cause messy drips.

Holding the squeegee at a 45° angle to the table, pull the resist across the screen in a smooth, even movement.

Watch the screen as you pull the squeegee. All the pattern areas should now look saturated with resist. Setting aside the squeegee, lift up the

Top left: Aligning the test tile under the screen

Top right: Taping the L in place

Bottom left: Loading resist on the screen

Bottom right: Pull squeegee toward you at a 45° angle.

Screen-printed tiles should dry for a few
hours before being glazed.

screen and examine the test print. If the print is complete and uniform, proceed to pull prints directly on tile. Otherwise, identify the issue (see "Troubleshooting," later in this chapter) and pull a few more test prints until you feel confident to print directly on bisque.

Screen printing is a skill. Initially you may pull incomplete prints, load the screen with excess resist, or make any of a large number of messy beginner errors. But within a short while, you'll be pulling crisp and beautiful prints tile after tile.

Cleaning and Storing

Screens must be cleaned immediately after use, since resist that dries on the screen is almost impossible to clean off.

Clean screens outdoors or in a well-ventilated area. Use a cotton rag to wipe away as much resist as you can, working first around all four edges of the screen and then toward the center. When all the surplus resist has been wiped off, soak a rag in screen-cleaning solution and scrub the screen thoroughly. Hold the screen up to the light to make sure no resist remains. Then allow it to dry in a vertical position, away from direct sun, before storing it away.

Troubleshooting

Incomplete prints can result from the squeegee being pulled unevenly or from insufficient resist on the screen. If the imperfection is minor, lower the screen back gently, angle the squeegee, and pull resist over the imperfect area only. This will usually correct the issue but may occasionally result in an unsightly double print.

Double-printed tiles have to be discarded. If you'd rather not risk a second pull, load a paintbrush with resist and touch up the imperfections manually.

If you notice the resist forming powdery streaks across the screen as you pull your squeegee across, the resist formulation is too dry. Add a few drops of linseed oil to the resist paste to correct this issue.

Printing Cuerda Seca Lines on Vertical Surfaces

Cuerda seca is a tile-glazing technique, done on horizontal surfaces. Once in a rare while, it has been attempted on vertical surfaces such as the large pots on the grounds of the Adamson House in California.

On vertical surfaces, lines cannot be screen-printed by the method previously described. They need to be hand-painted or stenciled, and the glazes cannot run even a little bit. Before you attempt cuerda seca on vertical surfaces, check the flow properties of your glazes by firing a few glazed tiles vertically.

Above: An incomplete print

Cuerda seca planter at the Adamson House, California. *Diana Mausser*

SECTION 6

Cuerda Seca Glazes

Cuerda seca differs from other well-known tile-glazing techniques in that pigments—oxides or fritted stains—are mixed into glaze before it is applied on tile. This direct application of "colored" glazes gives cuerda seca its distinctive vibrant appeal.

Because so many colorants are stable at earthenware temperatures, it is fairly straightforward to develop a glaze palette for cuerda seca. To keep it simple, you could start with off-the-shelf liquid glazes and step it up to in-house glaze formulation. Here we'll examine glaze options, with the pros and cons of each.

A comparison of surface-glazing techniques

Using Ready-Made Glazes

Liquid earthenware glazes formulated to mature around cone 06 are easy to use, come in many colors, and are consistent from batch to batch. On the flip side they are expensive, especially when shipped in from far away. However, since a little goes a long way, they remain a viable option for the small studio.

Formulated mostly for brushing on, ready-made liquid glazes are viscous and must be watered down slightly for cuerda seca application. Opaque glazes, rather than transparent or translucent glazes, provide excellent coverage even after dilution. In the United States, ceramic supply stores offer a range of liquid glazes in trial sizes and larger pint bottles. At the time of writing, there is not a consistent supply of liquid glazes in India.

Many ceramic supply stores affix indicator tiles to glaze shelves, to show how the glaze will look when it is fired. These tiles are usually made from white clay. Remember that the same glaze will appear darker and deeper on a red clay body.

While liquid glazes offer unmatched ease of use, there are instances when they are not the best choice, even for a small studio. Colors may suddenly get discontinued. The glazes may craze heavily on your clay body. A client may request tile in a custom hue. Therefore, for better control over your glaze palette, you may wish to formulate your own glazes.

Above: Flameback Tile's cone 06 glaze palette

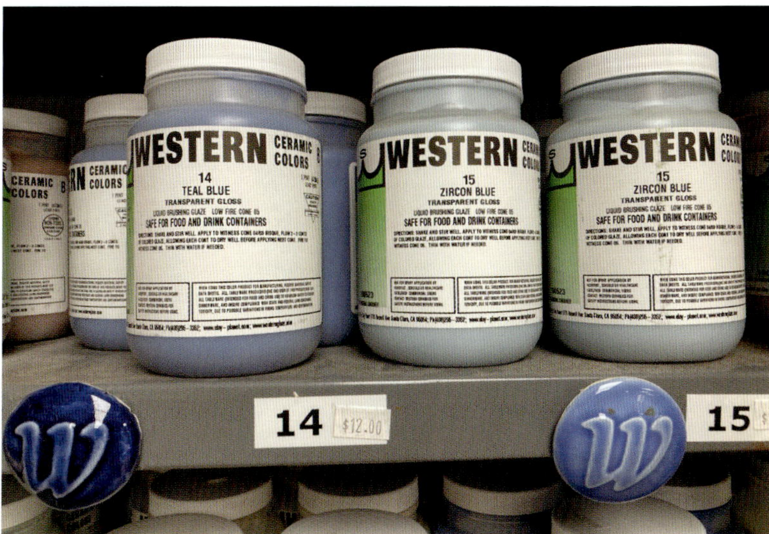

Above: Liquid glazes at Clay Planet, San Jose, California. *Suhita Shirodkar*

Left: Sample tiles indicate fired color. *Suhita Shirodkar*

Using a Commercial Base Glaze

A midway option between ready-made glazes and complete formulation is to purchase a commercial base glaze, test it for compatibility with your clay body, and then customize the color. This allows you to replicate glazes that were discontinued or to create tints and shades not available ready-made.

The optical properties of glazes differ, even when they mature at the same temperature. Therefore, the base glaze for a project should be the same, with only the colorants varying. For example, a cone 06 satin matte base and a cone 06 opaque gloss base on the same project are not advisable, unless you are aiming for a specific visual effect.

Colorants added to a base glaze can be oxides or stains. Stains are marvels of the ceramic industry. They contain an oxide or a blend of oxides in a fritted base. Because stains have already been fired much above earthenware temperatures, their color response is predictable and consistent. Some stains react unpredictably when the base glaze contains zinc. Check the oxide analysis of the base glaze to confirm the absence of zinc.

Above left: Mason stains line the shelves at Clay Planet, San Jose, California.
Suhita Shirodkar

Above right: A reference chart for Mason stains, with 12% stain added to an opaque base glaze. Clay Planet, San Jose, California.
Suhita Shirodkar

The higher the percentage of stain, the deeper the color, up to a certain point. Since the optimal percentage varies by stain, you should run line blends of base glaze and stain (or oxide), starting with about 4% stain and up to about 12%. Some stains can even be mixed like paints to create intermediate colors.

If the base glaze is clear, an opacifier must be added. A rule of thumb is 100 parts of glaze with 15 parts of colorant + opacifier. So, for example:

BASE GLAZE	100 g
ZIRCON AS OPACIFIER	9 g
STAIN	6 g

OR:

BASE GLAZE	100 g
TIN OXIDE AS OPACIFIER	7 g
STAIN	8 g

Approximately 0.5% CMC must be added for smooth flow through a bulb or precision-tip applicator.

Making Your Own Glaze

At earthenware temperatures, a lead or borate frit can function as a complete glaze. Ferro frits are used here, but you can test local equivalents that satisfy the same oxide formula.

Ferro Frit 3124-2, which contains about 14% boron, is an example of a near-complete earthenware glaze. Some ball clay helps suspend the frit, and bentonite keeps it from settling rapidly. A basic formulation looks like this:

FERRO FRIT 3124-2	85 g
TIN OXIDE AS OPACIFIER	5 g
BALL CLAY	8 g
BENTONITE	2 g

Zircon is less expensive than tin oxide and less likely to induce crazing. However, it produces a harsher white than does tin oxide. A formulation with zircon reads like this:

FERRO FRIT 3124-2	80 g
ZIRCON AS OPACIFIER	10 g
BALL CLAY	8 g
BENTONITE	2 g

To further reduce the likelihood of crazing, some or all of the Ferro Frit 3124-2 may be replaced by Ferro Frit 3249, a low-expansion magnesium borosilicate frit. Approximately 0.5% CMC must be added for smooth flow through a bulb or precision-tip applicator.

Adjusting Glaze Density

Cuerda seca glaze must flow smoothly out of its applicator—either a bulb or a precision tip—and dry on the tile in the correct thickness. The thickness of the glaze layer should be that of a playing card or a business card viewed on edge.

Too thin a glaze layer means the tile will lack depth and vibrancy; too thick an application wastes glaze and increases the chances of pinholing. The thickness of the glaze layer depends on the artist's skill and on the density of the glaze. Glaze density must be adjusted before glaze is applied.

To adjust glaze density for cuerda seca, a two-pronged approach—using your senses and measuring specific gravity—reduces the chances

The desired thickness of a cuerda seca glaze layer is about as much as a playing card. At righ you see a playing card set on edge.

Left: Glaze applied too thick, with unfortunate color choices, earmarks this tile for the scrap heap.

of error. Specific gravity compares the weight of glaze with the weight of the same volume of water.

Because bottled glazes are formulated for brushing, they are often viscous, with a consistency like paste. They must be diluted with water to make them flow out of a bulb or precision tip, but it is easy to go too far with the dilution. This two-pronged approach to adjusting glaze density minimizes the chances of overdilution.

Testing with Your Senses

Stir the glaze vigorously and then abruptly stop stirring. The liquid should remain in motion for a few seconds. If it becomes still immediately, it is too dense.

Pour approximately 150 mL of glaze into a small bowl. Observe how it pours. It should have the pouring consistency of cream. If it pours like skim milk, it is not dense enough.

Dry your hands thoroughly and dip your index finger into the bowl of glaze. Observe how it coats your finger. If you cannot see the outline of your fingernail, glaze is too dense.

If the glaze does not coat your finger evenly but slides off leaving only a patchy, watery film, it is not dense enough.

Glaze of the right density coats your finger completely but allows you to discern the outline of your fingernail. This is the right point at which to measure and record its specific gravity.

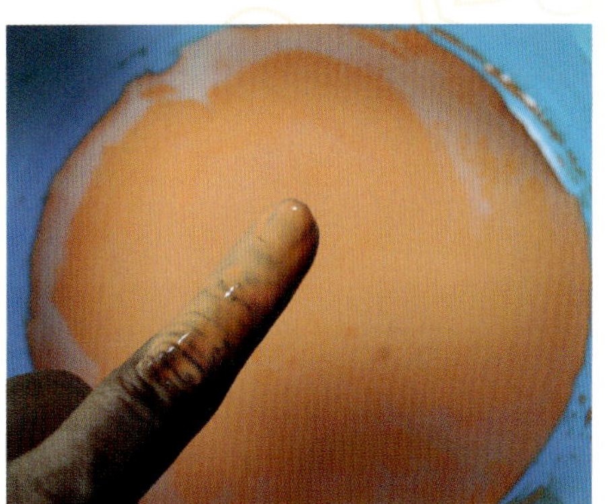

Top to bottom:

Fingernail not visible under layer of glaze. Glaze is too dense.

Finger and nail are clearly visible under the glaze. Glaze is not dense enough.

The right density

Measuring Specific Gravity

Specific gravity compares the weight of 100 mL of glaze to that of 100 mL water, which, for all practical purposes, weighs 100 g.

On a digital weigh scale, tare a 100 mL graduated cylinder, pour in 100 mL of well-stirred glaze, and record the weight.

If 100 mL of glaze weighs 150 g, then its specific gravity is 150/100 = 1.5, which means the glaze is one and a half times denser than water. Record this specific gravity. If this glaze gives you good results in the firing, you know that its target specific gravity is 1.5. This is the number to aim for the next time you mix up this glaze.

Target specific gravity varies by glaze. For cuerda seca glazes, it is usually between 1.35 and 1.55.

Specific gravity can be read with a hydrometer as well. Stir the glaze thoroughly, float the hydrometer in it, and take a reading off its side.

Measuring specific gravity with a hydrometer

Crazing

Glaze fits best when it is under slight compression with respect to its clay substrate. At earthenware temperatures, though, it is more likely that the glaze is under tension. Glaze under tension tends to craze.

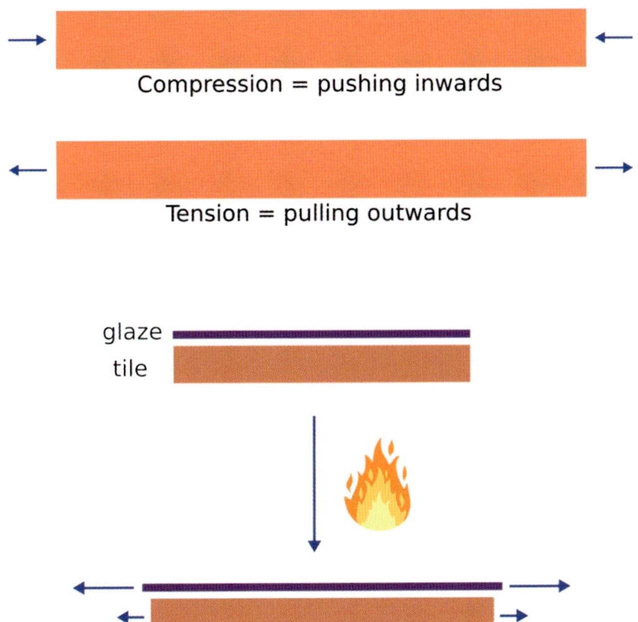

Glaze expands more in the firing than the underlying tile, crazing as it cools.

Crazing, seen as hairline cracks across the glaze surface, usually occurs while tile is still in the kiln, but can occur even days or weeks later.

If the tile will not be used in wet areas, crazing can be used for decorative effect. Craze lines can be accentuated by rubbing in dark oxide before sealing the tile.

Accelerated Testing for Crazing

The freeze-boil test is designed to accelerate conditions that lead to crazing. It stresses glaze by alternately subjecting it to boiling and freezing water.

Fill a large pot with water and bring it to a rolling boil. There should be enough water to submerge the tiles. Fill another large pot with ice water and set it close by.

Immerse test tiles in boiling water for three minutes. Remove them with tongs and immediately transfer them to the ice water, also for three minutes. Repeat the boil-freeze process for two more cycles.

Observe the glaze for crazing. If there is no visible crazing, dry the tiles and rub them with a dark oxide to detect microcrazing. The finer the crazing, the more difficult it is to reduce or eliminate. Here are a few approaches for reducing crazing:

1. Formulate a fritted glaze in which the frit has a very low coefficient of expansion.

2. Modify the clay body formulation to increase its coefficient of expansion.

3. Swap opacifiers, from tin oxide to zircon for example.

4. Increase the final soak time in the glaze-firing cycle, watching cones carefully to prevent overfiring.

5. Soak before peak temperature is reached.

6. Slow down the cooling. Do not open the kiln until it is below 100°C. Leave tiles in the kiln until they cool completely.

Glazing

Glazing cuerda seca tile is slow, steady work. A workbench with natural light and ergonomic seating is important. It helps to take breaks, to stretch, and to walk around at intervals. In a well-lit and comfortable setup, cuerda seca glazing is a meditative, "color-by-numbers" experience reminiscent of childhood.

Before you start glazing, check each screen-printed tile to ensure all lines (except textural lines) are closed. Correct imperfections with a paintbrush and resist, and let tile dry for a half hour before glazing. If you're using precision-tip applicators, fill the applicators and insert a pin into each tip to prevent glaze from drying in the cannula.

Precision-tip applicators ready for use

Quick dip in water before glazing gives
even results.

Best Practices for Glazing

1. Always handle a tile by its edges and corners; never touch
   its face.

2. Complete multiple-tile projects within a few sittings,
   without large gaps of time in between.

3. As far as possible, fire a project in a single firing, keeping
   similar colors on the same kiln shelf.

4. For a smooth and even coat of glaze, quickly dip the tile in
   water before you start glazing. Of course, you can do this
   only once, before you apply the first color.

5. If glaze accidentally jumps a line, do not try to clean it
   immediately. Let it dry first and then scrape off the excess
   with a pin tool or scalpel.

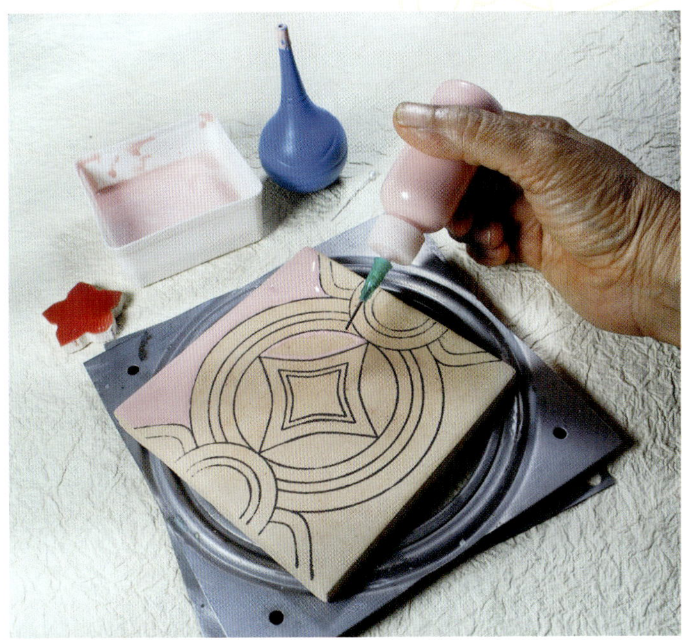

Glazing Techniques

Glaze with a sure, smooth hand. Use a bulb for large areas and a precision tip for smaller areas.

Precision tips come in many gauges. For very small areas, use the finest tip that will permit glaze flow. When you swap a tip, soak the used tip in a bowl of water until you wash it clean. This ensures that glaze does not block the cannula.

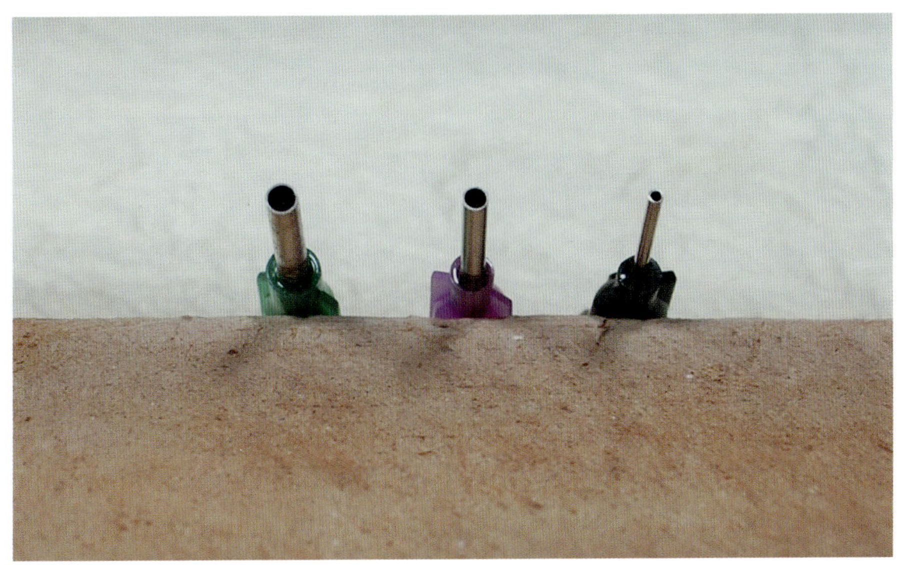

Top left: Bulb glazing quickly covers large areas.

Top right: A precision-tip applicator ensures neat application.

Left: Swap to a finer tip to glaze smaller areas.

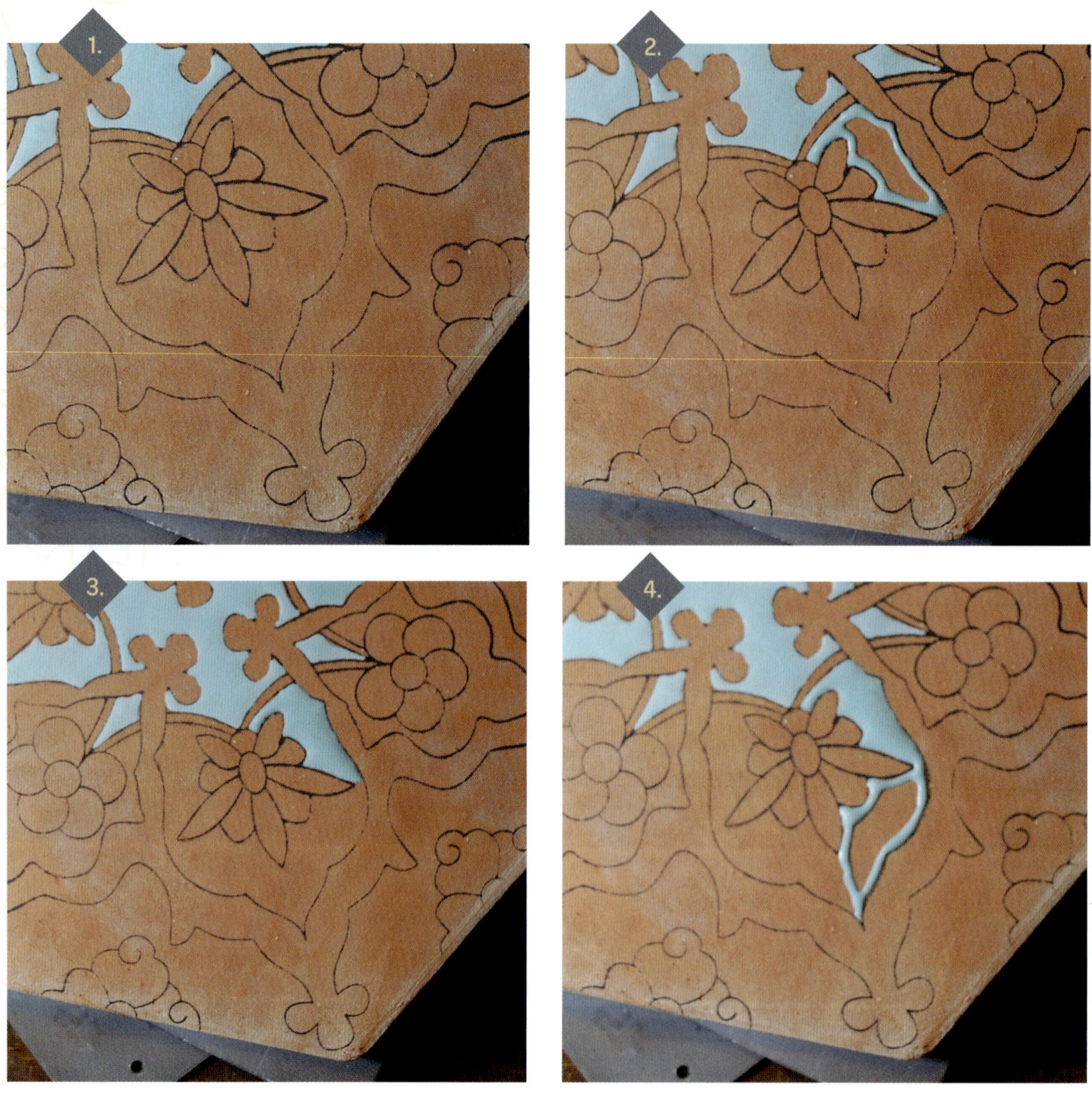

The Outline-and-Pool Method

The outline-and-pool glazing technique gives smooth, even coverage. It's important to glaze contiguously, so that glaze has an opportunity to settle down in an even layer as it dries. Since you are applying glaze and not pigments, you need not control the direction of your stroke as you would with majolica. Glaze will even out when it melts, and stroke direction will not be visible.

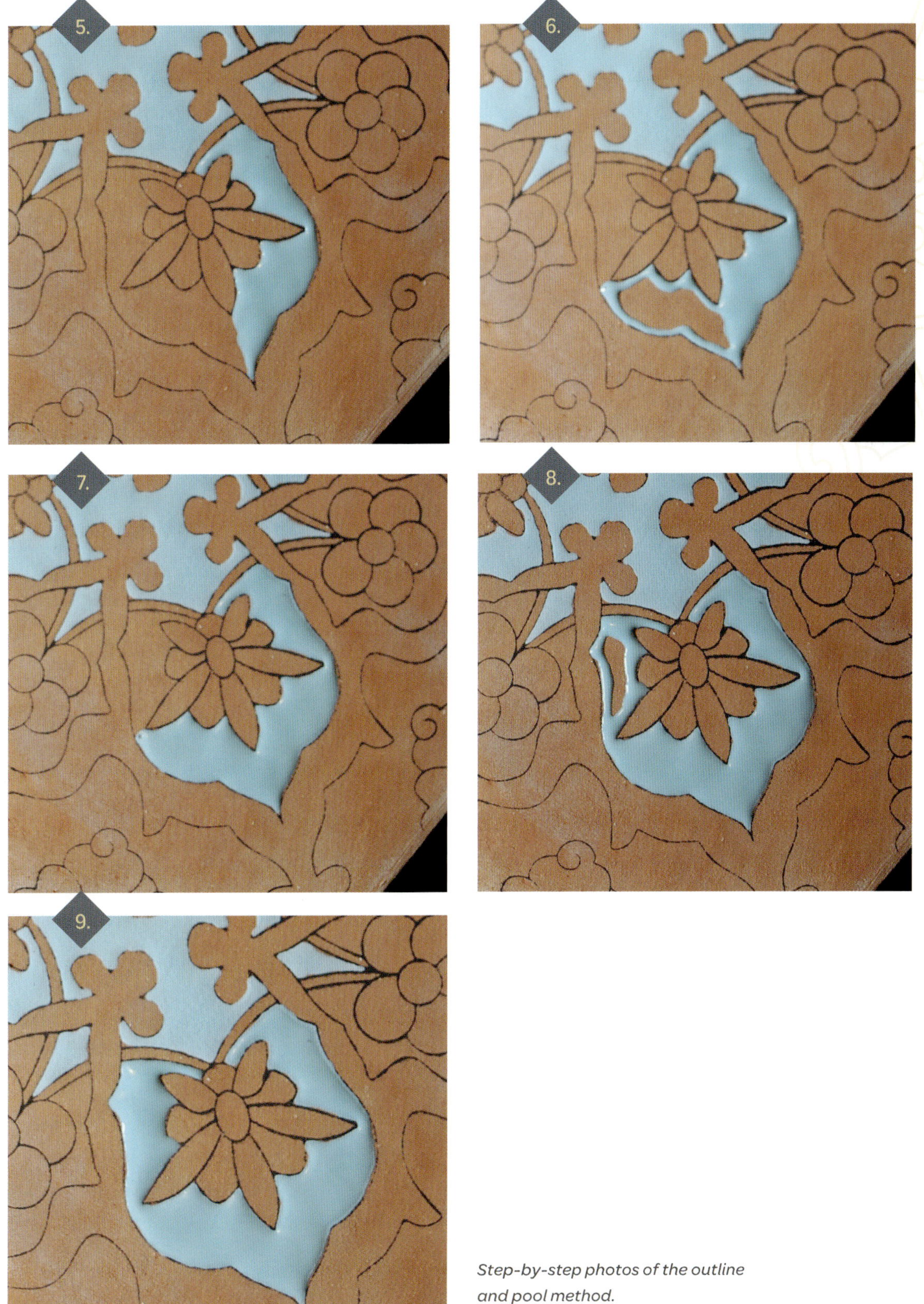

*Step-by-step photos of the outline
and pool method.*

In majolica, brush direction remains visible in the final work, but not so in cuerda seca. Detail from a majolica mural in the Menezes Braganza Hall, Panjim, Goa, depicting Vasco da Gama setting sail for India.

"Retract the Bead"

Do not end your glaze stroke in a narrow, closed area of the pattern. This will cause extra glaze to be deposited in a constrained area, where it has no opportunity to spread when it melts in the kiln. Melted glaze will cool into an unsightly bead or ridge. Always pull the glaze back into an open area, where it will not bead up.

Area of detail

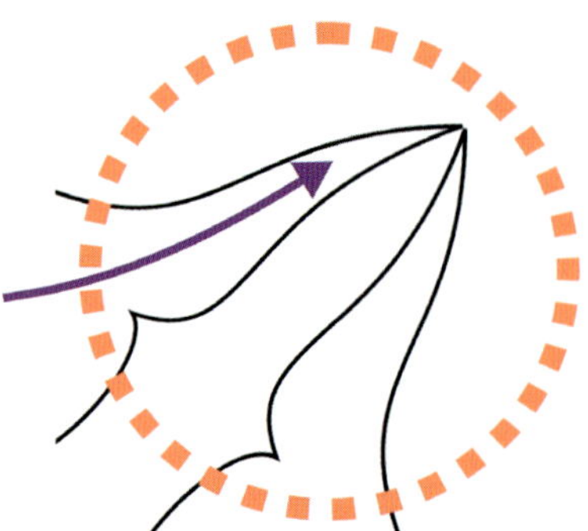
Incomplete stroke. Glaze will bead up at tip.

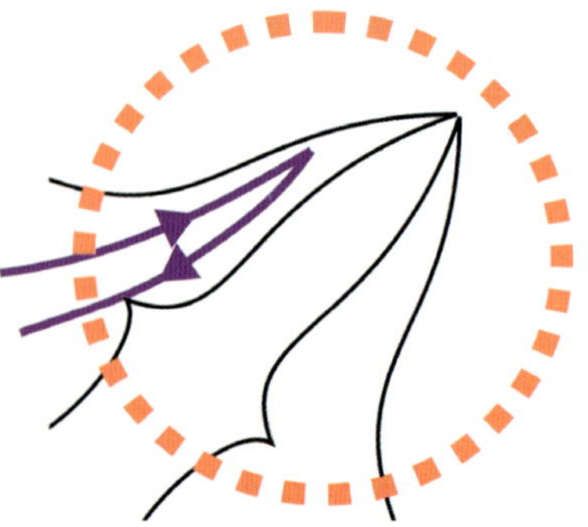
Correct stroke. Excess glaze pulled back from tip.

Glazing the Smallest Areas

To glaze areas too fine for the smallest precision tip, use a needle or pin to pull the glaze in. This should be an occasional requirement only, since glaze tends to retreat from tiny areas during firing. If you find too many such tiny areas on a tile, modify the pattern.

Finishing

After glaze has dried on the tile, correct any glaze jumps with a pin tool or scalpel. Too many jumps indicate that the glaze is too watery or that the pattern must be modified to have wider resist lines.

Top left: Glazing the smallest areas

Top right: Finishing

Glaze Firing

Either an electric or a gas kiln can be used for firing, but electric will yield a more consistent color response within a firing and across firings as well. Nevertheless, some color variation is inevitable, and expectations must be set accordingly, especially for custom projects or when color-matching is important.

Tips for Loading the Glaze Kiln

1. Tiles must not touch each other.
2. If a color tends to show variation, load all tiles of that color on one shelf.
3. Maintain a couple of inches headroom between kiln shelves.
4. If the kiln is lined internally with ceramic fiber blanket, flecks of blanket can dislodge and settle on tiles during firing. As a precaution, leave the top shelf in the kiln empty, to serve as a shield for the tiles below.
5. Ensure adequate clearances everywhere. Standard clearances for a midsized kiln are 2" from the lid, 1" from the kiln walls, at least 1" from the bottom, and 2" from the thermocouple.
6. Vacuum the kiln interior every few firings.
7. Place cones in the kiln. Cones tell the story of the firing better than a controller.

The Glaze-Firing Cycle

Although there are as many glaze-firing cycles as there are ceramists, a few general principles hold. For example, a glaze firing is faster than a bisque firing, since there is no moisture to vent. In a glaze firing, you need not have multiple midcycle soaks, although a soak near the end is useful to ensure color consistency. Some artists prefer to soak at the peak temperature; others take the temperature to peak, bring it down about 50°C (or 100°F), and then soak until the cones drop. In my Skutt, I soak at 950°C for twenty minutes before climbing the final ramp to 996°C. Watch the cones and keep diligent notes during the first few firings, to arrive at a firing cycle that works best for your studio.

Here is a sample glaze-firing cycle, with a total firing time of about seven hours. The slight variation between the tabulated firing cycles in °C and °F is to make it easy to program the kiln controller. It will not affect the results of the firing.

When color consistency is a must, load all tiles with similar colors on one shelf.

FIRING CYCLE (TEMPERATURES IN °C)

RAMP AND SOAK	TARGET TEMPERATURE	TIME	NOTES
Ramp at 100°C/hour	150°C	About 1.5 hours, depending on the outside temperature when firing began.	Lid is down and all peeps are closed.
Soak	Steady at 150°C	10 minutes	Brings the entire load up to temperature.
Ramp at 200°C/hour	950°C	4 hours	
Soak	Steady at 950°C	20 minutes	Brings the entire load up to temperature before the final ramp.
Final ramp at 100°C/hour	996°C	~ 0.5 hour	
Final soak	Steady at 996°C	0.5 hour, or until cone indicates end of cycle.	

AN EQUIVALENT FIRING CYCLE IN °F

RAMP AND SOAK	TARGET TEMPERATURE	TIME	NOTES
Ramp at 200°F/hour	300°F	About 1.5 hours, depending on the outside temperature when firing began.	Lid is down and all peeps are closed.
Soak	Steady at 300°F	10 minutes	Brings the entire load up to temperature.
Ramp at 400°F/hour	1,750°F	~ 3.5 hours	At 550°F, if no moisture is escaping, close the top peephole.
Soak	Steady at 1,750°F	20 minutes	Brings the entire load up to temperature before the final ramp.
Ramp at 100°F/hour	1,825°F	45 minutes	
Final soak	Steady at 1,825°F	0.5 hour, or until cone indicates end of cycle.	

Cuerda seca lines burn off in the kiln.

Troubleshooting

ISSUE	SOLUTION
Gap between glaze and cuerda seca line	This is due to excess linseed oil in the cuerda seca resist formulation, causing it to seep beyond the resist line as it dries. Mix up a stiffer resist paste, with less linseed oil.
Glaze frequently jumps the line.	Modify the pattern so that lines are wider.
Resist lines burn off in the firing.	Increase the percentage of borate flux in the resist paste formulation. You could raise it from the current 10% to 15%, or even 20% if required.

SECTION 7

Photo: Diana Mausser

Design Considerations for Installation

Cuerda seca tile has exceptional ability to enliven spaces. When those spaces are private, such as in a bathroom or on a kitchen backsplash, the pleasure is shared by a few. Installed in public spaces—on a park bench or a library wall—cuerda seca tile comes into its own, a visual and tactile delight for an entire community.

Celtic mural in the author's front porch has pleased or perturbed neighbors for over twenty years.

Because handmade tile is not as thin, flat, or uniform as machine-made, installation requires special care. Commercial tile installations are subject to standards, such as the American National Standards Institute (ANSI) standards for the United States, which apply to the tile itself and to the materials and methods for installation. Although it may not be practical to adhere rigidly to such standards for individual handmade-tile projects, they are nevertheless useful as guidances.

The Installation Site

The structure of a wall and its width and location determine the success of a tile installation. Visual distance, which is the open space required in front of any artwork for it to be appreciated optimally, is also a factor.

Wall Composition

A wall on which tile will be installed must be strong and rigid, of a material such as brick, concrete, or sheetrock. Any surface layers of paint, plaster, or wallpaper must first be stripped off, so that tile adheres directly to the wall rather than to a lamina.

Because handmade tile is quite thick, you may think that the tile, once installed, will project excessively from the wall, and that the wall must first be sanded or ground down to inset the tile. This is unnecessary. Once installed and grouted, with grout beveled cleanly around the perimeter, the tile will appear almost flush with the wall.

Wall Dimensions

The width of a wall, more than its height, influences the visual impact of a tile mural. A tiny island of tile in a vast sea of wall is unlikely to lure viewers. To determine the appropriate wall width, we can apply a guideline that museums use when they display paintings. The guideline states that any artwork should occupy slightly over half the width of the wall on which it hangs. If a tile mural is X feet wide, then there should be no more than $\frac{3}{8}$ X of bare wall on either side of it.

Tile does not appear to protrude excessively after grouting.

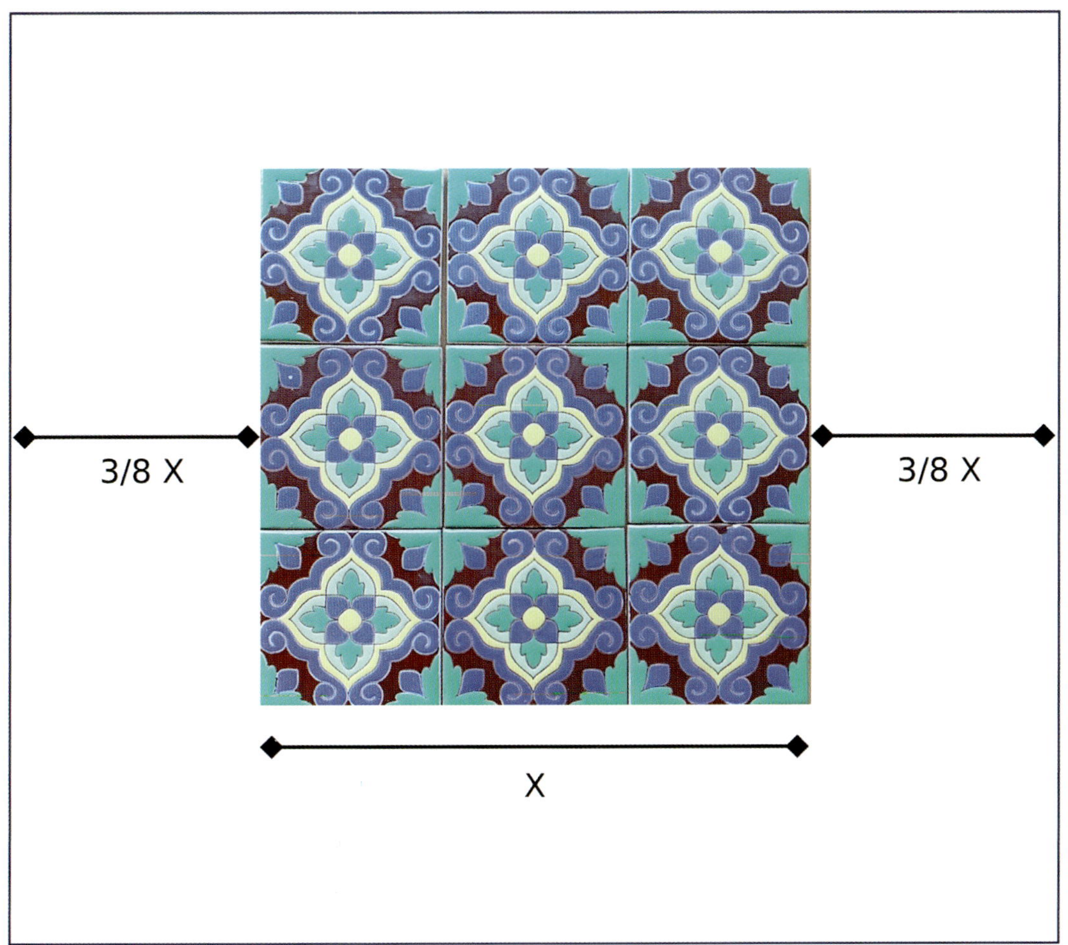

For example, for a mural 3 feet (36") wide, the wall space on either side should be

$$\tfrac{3}{8} \times 36" = 13.5"$$

And therefore, the entire wall should be no wider than
$$13.5" + 36" + 13.5" = 63", \text{ or } 5.25 \text{ feet.}$$

If two murals are to be installed as a pair, they must be treated as a single unit to calculate wall width. This raises the question of how much gap should be between them, since the viewer should recognize the two murals as distinct yet related. Again, a rule of thumb is that the gap between the two should be a fifth of their average width.

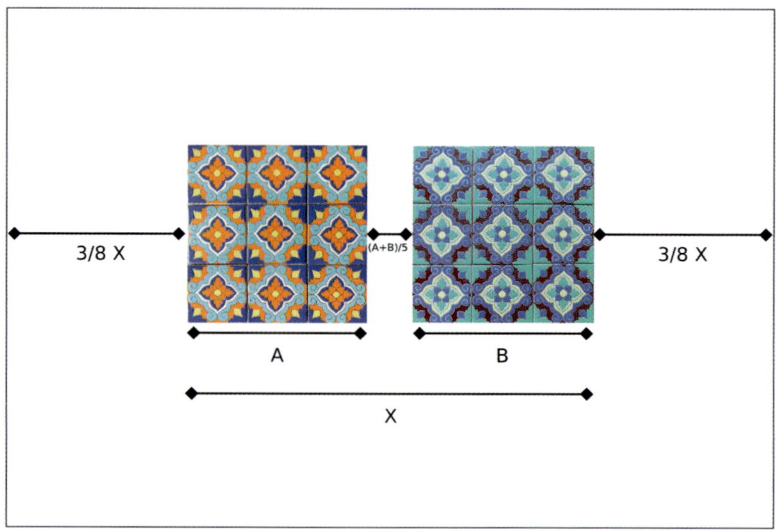

If two murals, each 36" wide, are to be installed side by side, then

Gap between murals = 36" / 5 = 7.2"

Total width of artwork (both murals treated as a single unit)
= 36" + 7.2" + 36" = 79.2". Wall on each side = ⅜ × 79.2" = 29.7".

Total width of wall = 29.7" + 79.2" + 29.7" = 138.6"

And therefore, for maximum impact, the wall should be no wider than 138.6", or about 11.5 feet.

While such calculations are useful, they should be treated as guidances only. Projections such as ducts, faucets, and electrical outlets need to be considered. The installer, who sees the wall in its context, must apply personal judgment as well.

Height from the Floor

The American museums standard for hanging artwork is 58" on center, which means the center of the installation should be 58" above the floor. For public installations, other standards may take precedence. For example, children's play areas or walls adjacent to access ramps may require lower installation.

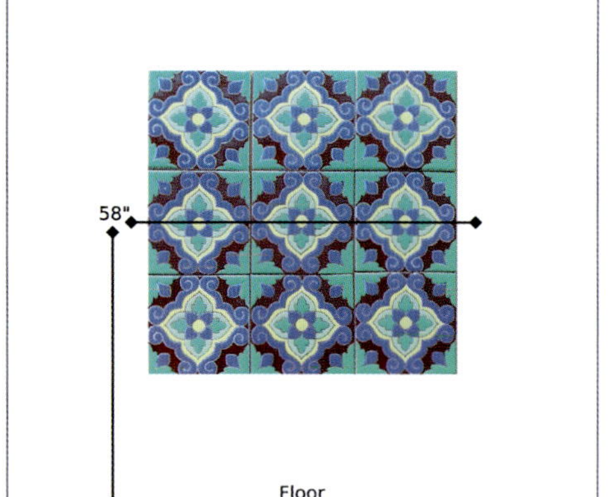

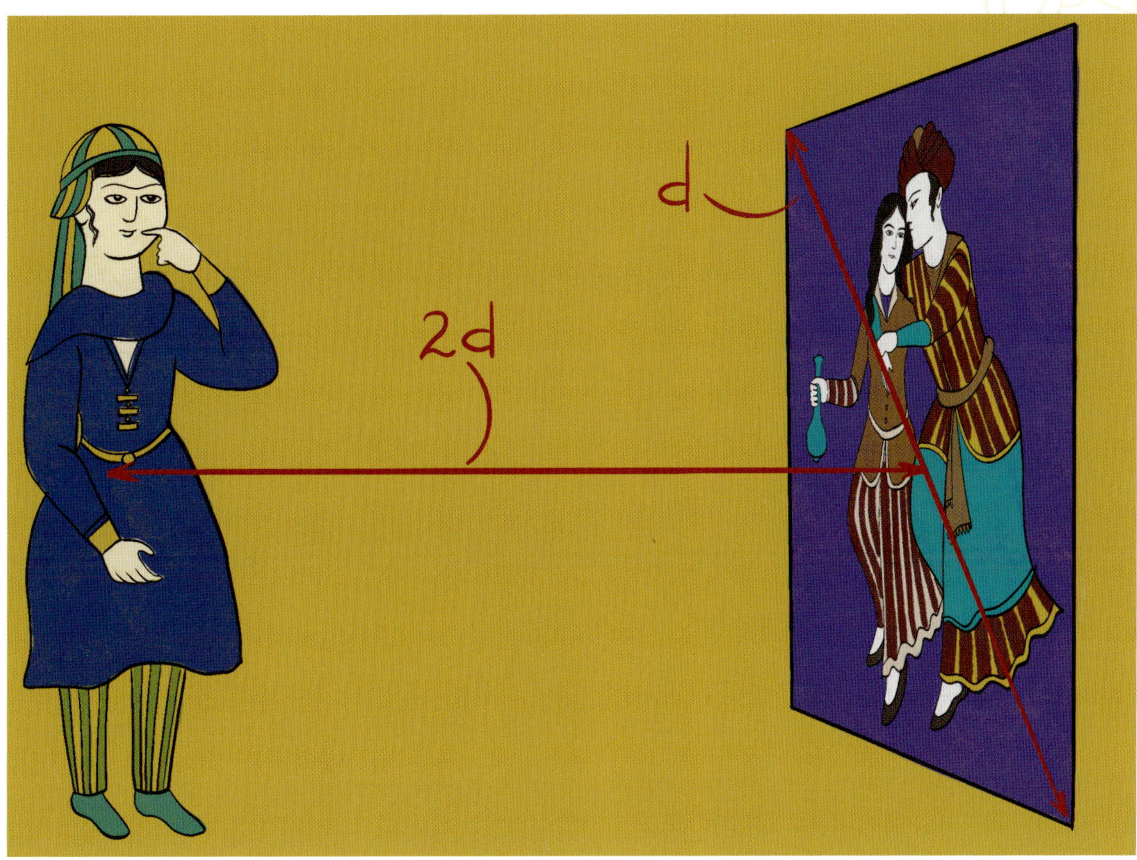

Viewing Distance

A viewer should be able to stand at a certain distance in front of a tile installation to appreciate it. What is that distance? Museum guidelines say that for any artwork to be viewed effortlessly, the viewing distance should be twice the diagonal length of the artwork.

By this guideline, for a square mural 36" wide:

$$2 \times (36 \times \sqrt{2}) = 101.8'' = 8.5''$$

There should be 8.5 feet of open space in front of the mural.

Viewing distance. *Mohini Bariya*

These tiles were sprayed with a strong jet of water. The sealed tile is at right.

Considerations for Outdoors and Wet Areas

You might have read that a clay body must be completely vitrified—and therefore nonporous—to withstand heavy rain or freezing temperatures. This does not explain how earthenware roof tiles in Japan and entire tile facades in Portugal have withstood centuries of rain or snow. Nor does it explain how large terracotta planters are successfully used for year-round landscaping in frost-prone regions of the world.

My experience with earthenware cuerda seca tile installed outdoors in Northern California (California Climate Zone 4) is that it can withstand decades of rain and subzero night temperatures. Cuerda seca tile that I have installed in shower stalls and around bathtubs has, at the time of writing, endured over fifteen years of daily exposure to direct water, with no ill effects.

One way, then, to think about earthenware is that it has the ability to absorb and exhaust moisture equally well. Cuerda seca tile can absorb moisture through the resist lines, which are unglazed, and along the edges, via the grout. If there is any crazing or microcrazing, moisture will travel across the tile face as well. But because the tile is semivitreous, moisture has an escape route and does not remain trapped inside the tile.

Nevertheless, it is a good idea to seal cuerda seca tile not once, but twice. Sealing before installation prevents grout from sticking to unglazed areas. A second sealing after installation seals the grout itself and ensures that most water hitting the tile simply runs off its surface.

Left: Tiles with crazing should be sealed well before installation.

Below: Cuerda seca Persian rug at the Adamson House, Malibu, California. *Diana Mausser*

Considerations for Furniture Installation

Since tile installed in furniture is viewed up close and routinely touched, it should be free of surface blemishes such as bubbling or crawling. Some crazing is acceptable as long as the tile is well sealed.

Considerations for Floor Installation

At the Adamson House in Malibu in Southern California is a marvelous "Persian carpet" of cuerda seca tile. The installation dates from the 1920s and is in very good condition. Although it is in the entryway, it is not clear how much foot traffic it has borne, since the Adamson House has been a museum—not a private home—for the past several decades.

A more modest floor installation is the strip of tile inset into the concrete walkway of my home (page 126). It has withstood years of foot traffic and of wheelbarrows being trundled over, with no visible impact. The clay body and tile thickness are identical to wall tile, as discussed in section 2.

These examples should provide some assurance about the durability of handmade cuerda seca tile set

Cuerda seca tile inset into a concrete walkway. *Brenda Marshall Barrett*

into floors. A conservative approach would be to install tile around the perimeter of a room rather than in central, high-traffic areas.

Grout: The Overlooked Design Element

Grout is a cement-based filler used between tile joints. Although often overlooked in the design process, it forms an integral part of a tile installation, completing and unifying the artwork. While several kinds of grout—sanded, unsanded, and epoxy—are available, sanded grout works best for handmade tile.

Grout comes in many colors. The safest choice is a neutral color, most often gray or off-white. For a bolder effect, grout can be either of a harmonizing or a contrasting color. Contrasting grout must be used with caution since it could end up dominating the installation.

Top: Neutral grout

Above left: Grout harmonizes with the wooden surround.

Above right: Contrasting grout

Installing into Furniture

Wood and ceramic tile complement each other. The tile can be functional, as on a tabletop, or decorative, as on this antique bench.

If you've never installed tile, a small furniture project is a good place to start. Here we walk through installing a set of nine tiles into a purpose-built wooden end table.

This set of nine tiles is slightly over a square foot, perfect for a small end table.

Antique bench with tile insets. Benches like this one, which can be suspended to swing gently, are traditional in Gujarat, western India.

Top: Setting up for furniture installation: plastic tub and sponge, disposable gloves, notched trowel, rubber float, scraper, wire brush, markers, tile spacers (optional), spirit level

Above: This basic table was purpose-built for the tile installation.

Setting Up for Furniture Installation

Preparing the Wood

Unfinished wood must first be sealed with a wood sealant, which prevents adhesive or grout from entering the grain of the wood. It also deepens the hue of the wood and imparts a gentle shine.

Preparing the Tile

Seal the tile with a tile sealant, to protect its surface and prevent grout from sticking to the cuerda seca lines. Place tiles in position to check for any fit or alignment issues. In this case, placing the tiles in the table inset shows that they sit low within the frame.

Whether tiles sit low or flush is a personal preference. Here, we'll attempt to set them flush because it makes the tabletop easier to wipe clean. There

are a couple of ways to do this. The first is to apply a thick layer of cement-based adhesive on the back of each tile. This boosts the tile but also makes the table heavier. A second option, used here, is to elevate the installation floor by gluing down a square of chipboard or plywood into the inset.

Top left: Tile sits low in its surround.

Top right: A square of ¼" chipboard glued to the table raises the installation surface.

Above: Adjust tiles to minimize alignment issues.

Setting the Tile

Once the chipboard has set, place tiles back into position. The chipboard has raised the floor so that the tiles are now virtually flush with the surround.

Rarely are handmade tiles perfectly even, and a few adjustments are inevitable. To adjust tiles into position, focus on the primary motifs or lines in the design and nudge the tiles around until their position pleases you.

Although there is not one perfect sequence in which the tiles should be affixed, for a project like this it helps to start at the center and spiral outward. With this method, tiles symmetrically occupy the available space, and grout width is equal on all sides. The fixing sequence for this table is as shown at left.

If glue is used for fixing, it should be a two-part epoxy adhesive, not instant glue or superglue. The longer setting time of epoxy adhesives permits time to adjust the positions of the tiles.

When the tiles are glued and their positions adjusted, cover the surface carefully with newspaper or a sheet of plastic. Gently lower a rigid board on the protective sheet and then distribute weights on top. Leave overnight.

Above: Sequence in which to fix the tiles.

Right: Tiles weighted down for a strong, permanent bond

Working grout into the crevices by hand

Grouting

Next morning, the table is ready for grouting. Follow the instructions on the grout package to mix up grout. About one cup of prepared grout will suffice for a project of this size. Although grout can be mixed with water, it is stronger and more flexible when mixed with a grout-bonding liquid.

You can use a grout float to push the grout diagonally into the crevices. For small projects like this, a gloved hand works beautifully, allowing you to work the grout down into the gaps.

Top left: Cleaning excess grout off the face of each tile soon after grouting makes subsequent cleanup much easier.

Bottom left: Wiping off grout haze

Right: The finished table, ready for use

Cleanup

Sanded grout dries quickly. Within a few minutes of application, excess grout can be wiped off with a damp sponge. This must be done gently so as not to gouge out the fresh grout.

Once the grout has dried fully (this may take up to an hour), wipe the haze off with a damp sponge. Multiple passes with the sponge are required to clear all the haze.

Then polish the table and let the polish dry completely before bringing the table into use.

CHAPTER 23

Installing on Walls

No matter how small the tile work, installing it on a wall requires planning and care. A vertical surface is always harder to work on than a horizontal surface. If the wall is external, exposure to the elements adds complexity. Here we'll walk through the installation of a nine-tile wall mural, slightly over one foot square. Applying the guidances discussed earlier, we know that the ideal wall width for this tile is about two feet, and the viewing distance is about three feet. We also know that the center of the mural should be 58" from the floor.

Mural for installation

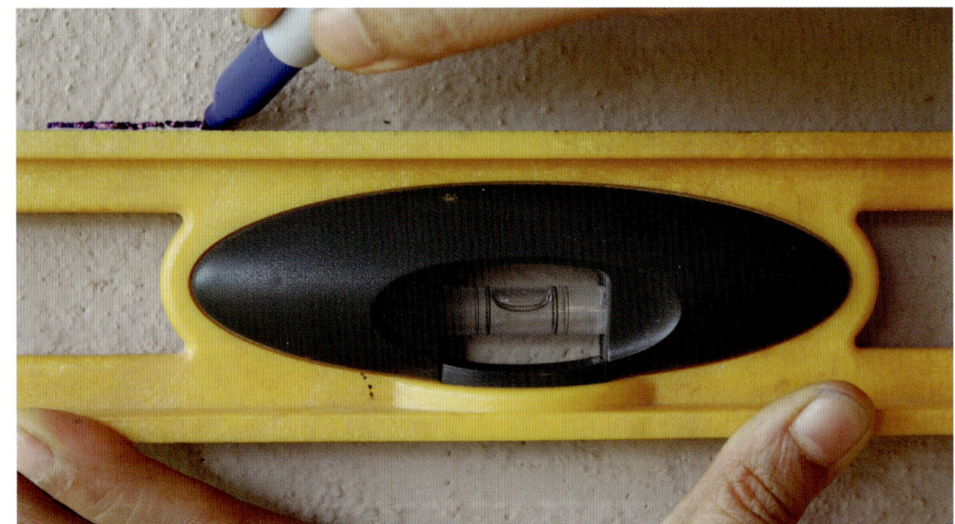

Top: Setting up for wall installation: plastic tub and sponge, painter's tape, disposable gloves, notched trowel, rubber float, scraper, wire brush, markers, tile spacers, spirit level

Center: Mark the starting line, using a spirit level for accuracy.

Bottom: Bottom row is complete.

What Kind of Adhesive?

Many types of tile adhesive are available. Some are pure polymers, while others are a mix of cement and polymer. For small projects like this, and if your tiles are quite flat, you could consider using a two-part epoxy adhesive. For larger projects thinset mortar, which contains cement and a bonding agent, works well. The advantage of thinset is that you can butter it on thicker wherever required, thus compensating for uneven or slightly warped tile.

Setting Up for Wall Installation

Site Selection and Prepping

Wipe down the wall and scrape off any peeling paint or plaster. Mark the vertical center of the wall and use your spirit level to draw a horizontal line marking the bottom of the mural.

Firmly tape a thin strip of wood along the line you just drew, to temporarily support tile during installation. Spread newspaper or plastic sheeting on the floor to catch drips. Seal the tile with sealant and set it down near you, ready to be installed.

Fixing the Tile

Follow the mixing instructions on the bag of thinset, paying particular attention to the setting time. Using a notched trowel, spread a thin layer of thinset on the area of wall to be covered by the bottom row of tile. Then apply thinset to the back of each tile and affix it in place.

While affixing, use a small push-and-twist motion of the wrist to ensure good contact between tile and wall. Tile may need to be held in place for a minute until it adheres.

Tape each completed row for temporary support. Watch the tiles as you work, making small adjustments wherever necessary. Insert tile spacers after each row.

Top: Spacers between tiles ensure even installation.

Bottom left: Tape across for temporary support.

Bottom right: Tile fixed and left to bond overnight

When all the tile is installed, tape diagonally and leave overnight.

Any thinset left over after installation should be disposed of carefully. Washing it down the sink can clog the drainage system.

Grouting

The next day, the mural is ready for grouting.

Follow the instructions on the grout package to mix up a small quantity, using grout-bonding liquid rather than water. About one cup of prepared grout will suffice for a mural of this size.

Use a grout float to push the grout diagonally into the crevices. Work grout into interstices first, and then along the perimeter.

Right: Tape peeled off, this mural is ready to be grouted.

Below: Work grout into the interstices (between the tiles) first.

Sanded grout dries quickly. Within a few minutes of application, excess grout can be wiped off. This must be done carefully so as not to gouge out the freshly laid grout. Once the grout has dried, wipe the haze off with a damp sponge. Multiple passes with the damp sponge are required to clear all the haze.

When the installation is dry, seal the tile again and touch up the paint around the mural if required.

Above: Grout haze wipes off with a damp sponge.

Right: The completed installation

Opposite Page

Top: Grout the perimeter after the interstices are grouted.

Bottom: Pay attention to corners.

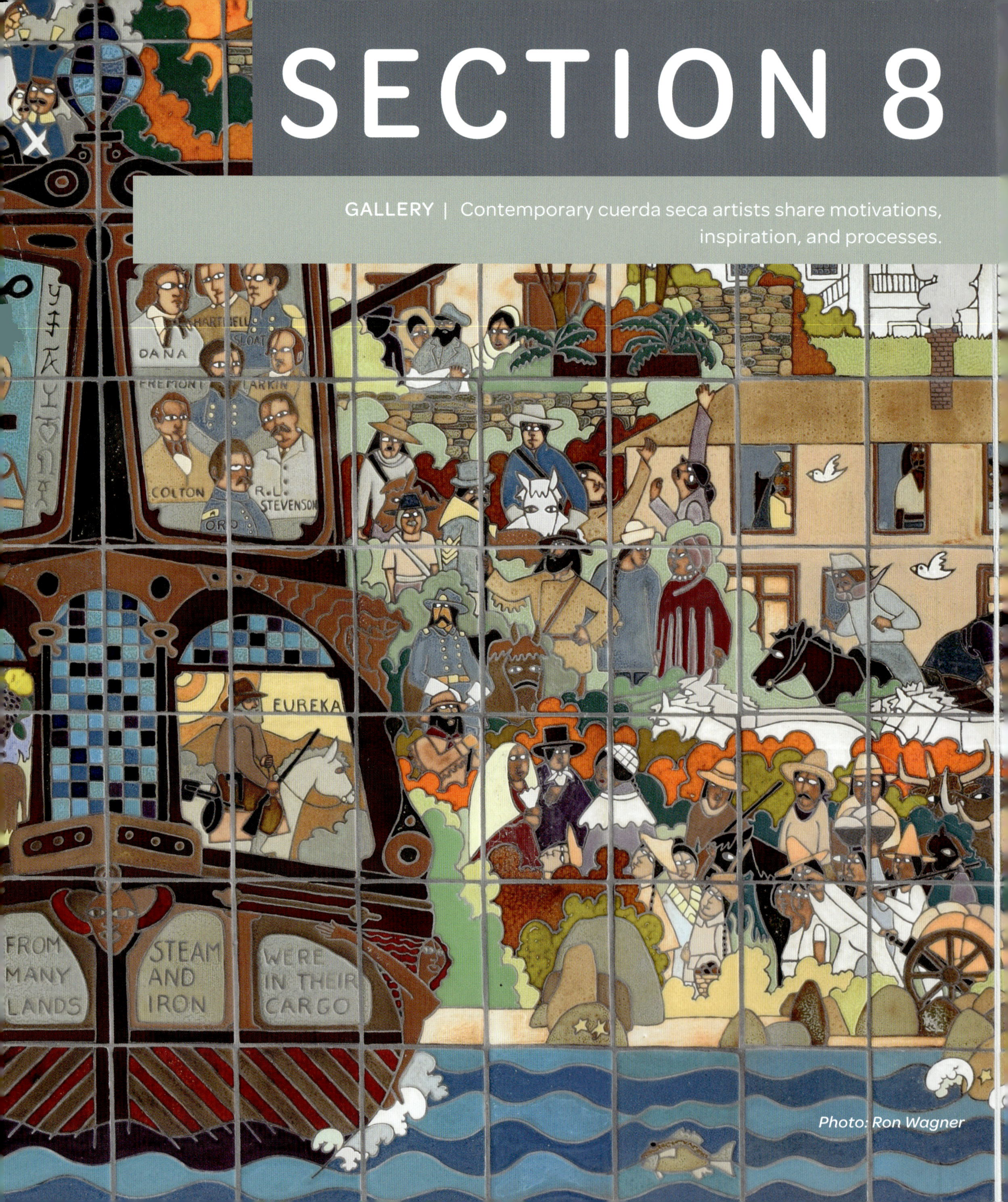

SECTION 8

GALLERY | Contemporary cuerda seca artists share motivations, inspiration, and processes.

Photo: Ron Wagner

Guillermo (Bill) Wagner Granizo

www.granizoart.com

Photo: Ron Wagner

San Francisco Bay Area muralist Guillermo (Bill) Wagner Granizo was a virtuoso of cuerda seca art. Bill came to cuerda seca in the '70s, after formal training in art, a career designing sets for television shows, and several years of work in ceramic mosaic. And he worked in cuerda seca, prolifically and masterfully, until he passed away in 1995.

Bill's 45-foot mural in downtown Monterey, California, exuberantly narrates the history of the city in over 150 charming vignettes. Created in 1984 and subsequently uninstalled and reinstalled, it retains the freshness and vitality of the day it emerged from the kiln.

For public commissions such as the Monterey mural, Bill researched meticulously but never drew plans on paper. On earthenware bisque purchased from Stonelight Tile in San Jose, California, Bill drew freehand

Bill's mural in downtown Monterey. *Ron Wagner*

designs with a felt-tip marker. Then he went over the lines with resist, "bulbed" on glazes, and fired the tiles in Stonelight's conveyer kiln. His son recalls that many vehicles were worn out from carrying heavy tiles back and forth. We do not know Bill's cuerda seca formulation, but quite likely it contained manganese dioxide suspended in a drying oil.

Bill was inspired by Mexican muralists including Diego Rivera, David Siqueiros, and Rufino Tamayo. You can see their influence in the distinctive way he renders faces in smoothly flowing resist lines.

Because Bill did not need mockups or sketches, he could work very fast. The Monterey mural, all 45 feet of it, took him only eight months from concept to installation. Probably his fastest project was a mural of over 2,000 square feet for the 1984 Los Angeles Olympics, which he completed within ten weeks.

Bill's murals invite the viewer to spend time with them and touch them. They hold large and small stories and then stories within other stories.

We are fortunate Bill chose to work in cuerda seca, so that some of his brilliant murals could grace public spaces for our viewing pleasure.

Details from the Monterey mural. *Ron Wagner*

Diana Mausser

Native Tile
www.nativetile.com

Native Tile, founded and led by Diana Mausser, was launched in a tumbledown Southern California boatyard shed in 1990. With a college degree from UCLA in ceramics and textile design, but no business experience, Diana took a leap of faith to set up an artisanal tile studio that might enable her to make a living doing what she loved. Thirty years later, Native Tile boasts an impressive client list, several industry awards, and a reputation extending far beyond its premises in Torrance, California.

Here is what aspiring cuerda seca artists can learn from Diana's experience at Native Tile.

Balance your time between the creative and business aspects: "I am the main designer and ceramist," says Diana, "but I run the business as well. I am happiest designing new patterns and working with clay, so I always struggle with balancing my two roles. It has been a learn-as-I-go approach, which has been difficult."

Blocking out chunks of time during each workweek for handling marketing, sales, and accounts is one way to ensure that you keep an eye on the business while handling the daily technical and production challenges of a tile studio.

Know when and how to bring in technology: "Another challenge has been incorporating appropriate new technology into the business in ways that can be helpful, without getting caught up in the technology-driven world. Our handmade processes make our tile unique and soulful, and we plan to keep it that way, but I am aware that we need to use technology to help us stay visible and apparent."

On the production side, Native Tile has a hybrid design model, in which hand-drawn patterns are digitized for making screens and building a pattern library. On the business side, Native Tile has embraced technology with a powerful web and social media presence that instills confidence in the product and the business. And yet, the tile itself stands testimony to technical skill and old-fashioned hand craftsmanship.

Diana Mausser

Detail from vintage Gladding McBean mural. *Diana Mausser*

Stay the course: "Keeping the business going through a severe economic downturn was challenging, especially when it coincided with a change in architecture to more-contemporary styles where patterned tiles were not as popular. We have had to weather a lot of changes and adapt without losing our style and integrity."

Native Tile's ability to stay the course is aided by strong relationships. Diana's emphasis on nurturing relationships with technical experts, showrooms, sales representatives, and employees has steadied Native Tile through economic downturns and shifting style trends. "Gather as many experienced people around you," advises Diana, "to create a good working team."

Embrace experimentation: "If you accept and embrace that experimentation and education will be a part of your ceramics journey, then you will enjoy the process tremendously."

Although Native Tile's production processes are reliable and time-tested, experimenting with glazes and clay bodies remains vital, especially on restoration projects. Once in a while, and this is most common with restoration projects where color matching is important, the color of the clay adversely affects glaze colors to such an extent that a new clay body must be created. "An example of this was when we created two tile murals for the historic Biltmore Hotel in Santa Barbara. [We wanted to] emulate the vintage Gladding McBean murals already on the property. The original glazes were semi transparent, which meant the color of the glaze was influenced by the clay underneath. The Gladding McBean murals used a unique buff clay body, which gave the lighter-colored glazes an underlying blush that you could not emulate exactly in the glaze itself. It had to come from the clay. So along with Sullivan Ceramics we developed the appropriate buff clay body that enhanced our glazes, bringing our murals closer to the original look that was so attractive in the vintage murals."

Here is the clay body recipe, arrived at after much experimenting, for the Biltmore Hotel project. In the recipe, dolomite serves a dual function, affecting the expansion properties of the body while also "bleaching" the iron oxide to achieve the desired shade of buff.

"D" LIKES BUFF (CONE 05-04): DEVELOPED BY SULLIVAN CERAMICS FOR NATIVE TILE

Greenstripe clay (this is a fireclay)	50 lbs.
Dolomite	25 lbs.
F-70 silica sand	20 lbs.
F-3269 frit (a lead-free, high-alkali, low-calcium frit)	15 lbs.
Silica (200 mesh)	10 lbs.
Barium carbonate	1.5 lbs.
Red Iron Oxide #4284	0.5 lb.
TOTAL BATCH WEIGHT	122 lbs.

Opposite page

Mural at the Biltmore Hotel, Santa Barbara, California. *Diana Mausser*

Detail from vintage Gladding McBean mural. *Diana Mausser*

RTK Studios

www.rtkstudios.com

Mary Kennedy and Richard Keit, RTK Studios

Santa Catalina Island, part of the Channel Islands archipelago, was developed in the late 1800s as an offshore resort for Southern Californians. In the 1920s, when rich seams of clay were discovered near the island's golf course, the Catalina Clay Products Company was born. Its vibrant cuerda seca tile was used to decorate the island's fountains, facades, and harbor. Decades later, a young Richard Keit spent summers on Santa Catalina and was charmed and intrigued by the bright, beautiful tile all around him. He began taking pencil rubbings of the tile, creating a design library. In some places, fragments of tile were used as mosaic, and that set Richard off on a treasure hunt across the island, trying to find the rest of the tile. So it was only natural that Richard would in 1979 set up RTK Studios, drawing inspiration from the cuerda seca tile that had delighted him in childhood.

By the late '70s, though, the world of ceramics had changed drastically. Many of the minerals and oxides used by the Catalina Clay Products Company and its contemporaries were recognized as toxic. Toyon Red, the company's unique orange-red glaze, contained uranium, now strictly out of bounds. Painstakingly, Richard began developing his own glaze palette. When Mary, with her background in visual arts and ceramics, joined RTK Studios in 1990, motivation and output multiplied exponentially. Today, RTK Studios has a vast design library including a trove of original patterns, a broad and impressive custom glaze palette, and happy clients across continents. Here Mary shares some of RTK Studios' business logic and experiences.

On whether it is necessary to stay current with design trends and styles: Although they stay aware of trends, Mary says the luxury of making their own designs and colors underpins their business model. "Most of our designs and all our glaze formulas are proprietary though influenced by the patterns and colors of history. Our stock tiles are not tied to the ebb and flow of the ever-changing market. Our tiles convey antiquity or modernity depending on how they're used in the installation. Our inventory fluctuates with new designs and color palettes we've created and with exciting developments in our glaze formulations. It keeps us in the creative mode of this art form."

On scaling the business up and down: Although the studio has expanded over the years, mostly because more space is required to store screens and inventory, it remains "low tech" at heart, and the kiln capacity has increased only slightly. "At one time we had twelve workers. We became less efficient and started losing control of quality, so it was best to [scale back and] keep things simpler in regard to the workforce."

On making a livelihood from cuerda seca: both Mary and Richard are motivated by the pleasure of making beautiful things that will last well beyond their own life spans. But they warn that "to make a financially sustainable career out of tile making is not an easy task. [It means] long hours, inconsistent materials and suppliers, equipment failures, utmost attention to all details, temperamental clients, and lots of good old-fashioned grunt work." Also, the cuerda seca process is geared to making multiples. "Therefore, it is not as easy to do one-offs as it would be with a direct-painting technique like majolica. It works best with bigger production runs."

Murals from RTK Studios' Porthole series, celebrating the island's rich undersea life, line the Enchanted Promenade in Avalon on Santa Catalina Island.

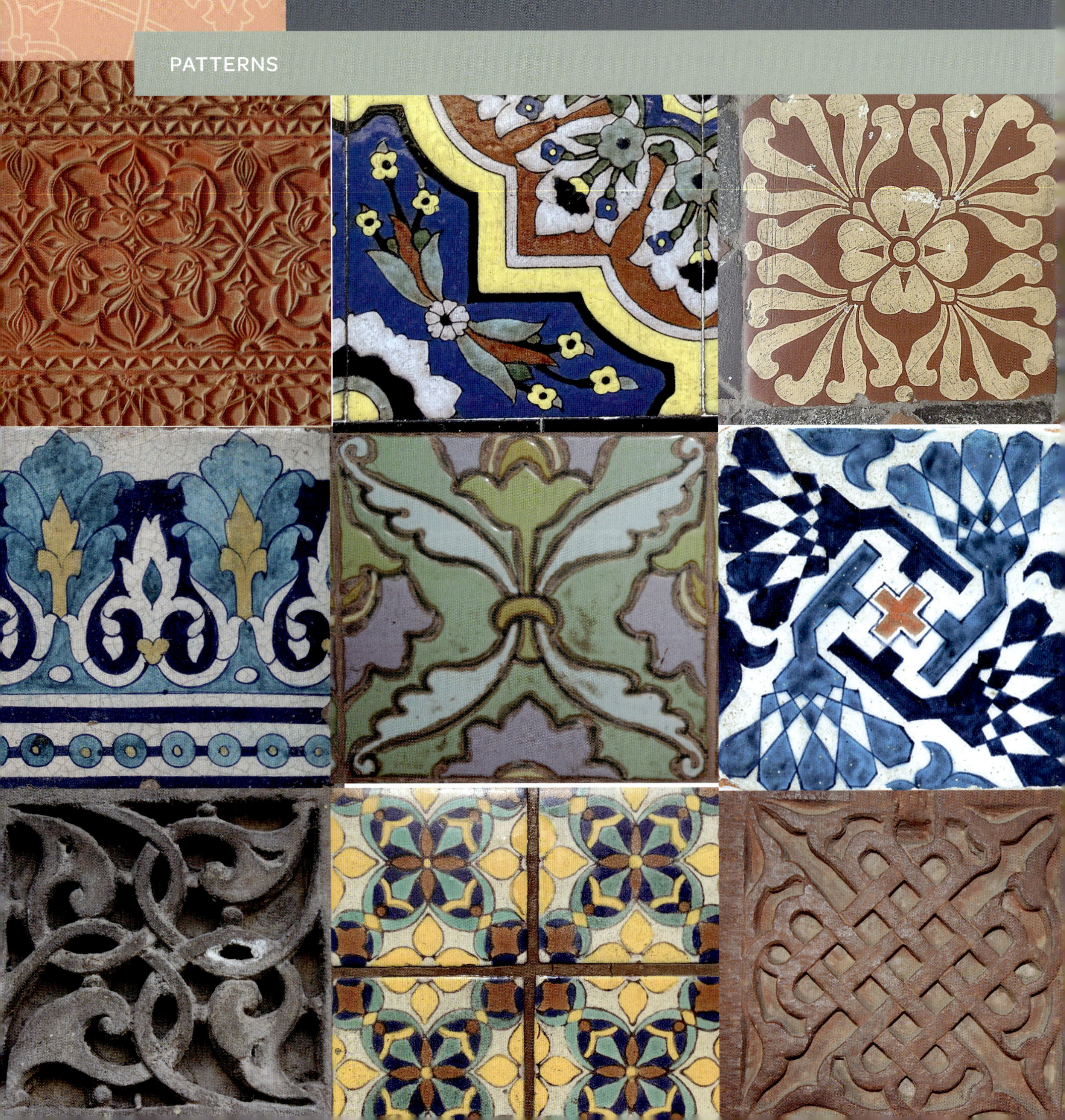

Patterns

The patterns shared here are our common heritage. Most are from medieval monuments in India, several of them UNESCO World Heritage Sites. A few are from modern structures or buildings in India and California. When you use these patterns, please acknowledge their provenance, as a tribute to the anonymous craftsmen who created them. Vector files for these patterns are at www.flameback-studio.com.

The hope is that you will use these patterns and get inspired to create your own, from historic buildings and architectural ornamentation around you. When you create patterns, especially when you translate from another medium such as stone or metal, try to translate the craftsman's intent rather than aiming for an exact reproduction. By doing so, you ensure that the constraints of the original medium or tools do not work their way into your cuerda seca pattern.

Alai Darwaza, Qutb Complex, Delhi, four-teenth century, and close-up of Alai Darwaza pattern. *Wiki Commons*

Champaner-Pavagadh complex,
Gujarat, ca. fifteenth century.
Wiki Commons

From the derelict remains of the Chisti Mahal, a sixteenth-century pleasure palace in Mandu, central India. These pictures were taken in 2010. Above, you see the ghost of the pattern, with a few fragments of ceramic revealing the original glaze colors. The pattern was high up on a ceiling, which has probably since collapsed.

Chisti Mahal with dome detail. *Wiki Commons*

Fatehpur Sikri, sixteenth century

This underglaze tile from Multan, ca. six-teenth century, was in a 2013 Sotheby's online auction.

Park bench in San Diego, California, twentieth century. At right is a digital tessellation.
Suhita Shirodkar

Menezes Braganza Hall, Goa, India. Floor-to-ceiling majolica murals surrounded by cuerda seca field tile in unusual glaze colors. Tiles made in Portugal. Early twentieth century.

NESTA FRESCURA TAL DESEMBARCAVAM
JÁ DAS NAOS OS SEGUNDOS ARGONAUTAS,
ONDE PELAS FLORESTAS SE DEIXAVAM
ANDAR AS BELLAS DEUSAS COMO INCAUTAS

Cement floor tile in Shiv Shanti
Bhuvan, one of Mumbai's iconic art
deco buildings, twentieth century.
Raj Mehta

Tile panel at an ice cream store in Alameda, California, twentieth century. *Mallika Bariya*

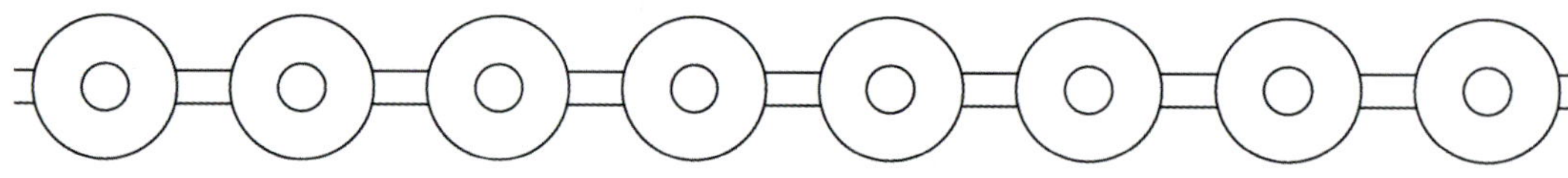

Underglaze tile purchased from the National Museum of Pakistan, Lahore. Probably early twentieth century, Multan.

SECTION 10

RESOURCES

Further Reading

Giorgini, Frank. Handmade Tiles. New York: Lark Books, 1994.

Lesch-Middleton, Forrest. Handmade Tile. Beverley, MA: Quarry Books, 2019.

Norton, F. H. Ceramics for the Artist Potter. Reading, MA: Addison-Wesley, 1956.

Porter, Venetia. Islamic Tiles. Northampton, MA: Interlink Books, 2005.

Rhodes, Daniel. Clay and Glazes for the Potter. Radnor, PA: Chilton Book, 1973.

Rindge, Ronald L. Ceramic Art of the Malibu Potteries, 1926–1932. Malibu, CA: Malibu Lagoon Museum, 1994.

Rosenthal, Lee. Catalina Tile. Sausalito, CA: Windgate, 1992.

Southwell, B.C. Making and Decorating Pottery Tiles. New York: Watson-Guptill, 1972.

References

These online resources were used frequently while writing this book:

Digitalfire, www.digitalfire.com

Glazy.org

NCECA blog http://blog.nceca.net/online-resources-for-ceramics

Introduction

O'Kane, Bernard. "Tiles of Many Hues: The Development of Iranian Cuerda Seca Tiles and the Transfer of Tilework Technology." in And Diverse Are Their Hues: Color in Islamic Art and Culture. Edited by Jonathan M. Bloom and Sheila S. Blair, 175–203. New Haven, CT: Yale University Press, 2011.

Section 1

Respirable crystalline silica, https://www.osha.gov/dsg/topics/silicacrystalline/

Workplace ergonomics, https://www.osha.gov

Section 2

Arbuckle, Linda. "Earthenware Clays." From her website https://www.lindaarbuckle.com/arbuckle_handouts.html.

British Geological Survey Mineral Planning Factsheets. https://www.bgs.ac.uk/mineralsuk/planning/mineralPlanningFactsheets.html.

Ceramicartsdaily.org. "Ceramic Raw Materials: Understanding Ceramic Glaze Materials and Clay Making." Newark, NJ: Ceramic Publications, 2012.

Clay Materials for the Self Reliant Potter. http://collections.infocollections.org/ukedu/en/d/Jg41cle/8.html.

Diedel, R., and S. Link. FGK Forschungsinstitut für Anorganische Werkstoffe–Glas/Keramik–GmbH, Hoehr-Grenzhausen, Germany. "Modulus of Rupture of Unfired Clays and Bodies." Paper presented at QualiCer 2006, IX World Congress on Ceramic Tile Quality, Castellón, Spain.

Dondi, Michele, Mariarosa Raimondo, and Chiara Zanelli. "Clays and Bodies for Ceramic Tiles: Reappraisal and Technological Classification." Applied Clay Science 96 (2014): 91–109.

Kaplan, Jonathan. "Techno File: Clay Body Building." Ceramics Monthly, May 2015. http://www. ceramicsmonthly.org.

Manfredini, Tiziano, Gian Carlo Pellacani, Paolo Pozzi, G. Marzola, and C. Pasquali. "Preparation of a Ceramic Floor Tile Body Containing Pure Bentonite as Strengthening Agent." Journal of the Spanish Ceramic and Glass Society 30, no. 1 (1991): 5–9.

Material Substitution in Clay Bodies. https://ceramicartsnetwork.org/ceramic-recipes/reference/material- substitutions-clay-bodies/#.

Rahaman, Mohamed N. Ceramic Processing. Boca Raton, FL: CRC Press, 2017. Plasticizing properties of sodium lignosulfonate.

"Talc and Asbestos." Ceramics Monthly, February 2008, 58–61.

Section 3

Flat Tiles the Easy Way. https://ceramicartsnetwork.org daily/pottery-making-techniques/making-ceramic- tile/flat-tiles-the-easy-way/.

Section 5

Toxicological Profile of Manganese. https://www.ncbi.nlm.nih.gov/books/NBK158871/

Section 6

Guidelines for Using Commercial Stains in Glazes. https://ceramicartsnetwork.org/daily/ceramic-glaze-recipes/glaze-chemistry/guidelines-for-using-commercial-stains-in-glazes/.

Laguna Clay Company. "Ferro Glaze Frit Data." http://www.lagunaclay.com/catalog/pdf/lcc_fritdata.pdf.

Mason Color Works FAQs. http://www.masoncolor.com/frequently-asked-questions.

Section 7

"Hanging Guide: Installing an Exhibition." Purdue University College of Liberal Arts. https://cla.purdue.edu.

How to Install Handmade Tile. http://handmadetileassociation.org. (This resource is no longer available, and the organization appears to have closed down.)

National Museums of Scotland. "Exhibitions for All: A Practical Guide for Designing Inclusive Exhibitions." https://www.rnib.org.uk.

Smith, David S. Email exchange; also "Techno File: Freeze-Thaw Myth." Ceramics Monthly, November 2016.

Walters, William, and Roy Harrison. "British Standards and Codes of Practice for the Installation of Wall & Floor Tiles." Submitted to Qualicer. http://www.qualicer.org/recopilatorio/ponencias/pdfs/9201060e.pdf.

Suppliers

There are many ceramics suppliers in the United States, several with both physical and online stores. To minimize shipping costs, find a supplier near you. In India, the number of suppliers is smaller but growing steadily each year. These are suppliers I have used and can recommend.

Ceramics Materials and Equipment

CLAY PLANET
1775 Russell Avenue
Santa Clara, CA 95054, USA
http://shop.clay-planet.com/

LAGUNA CLAY COMPANY
14400 Lomitas Avenue
City of Industry, CA 91746, USA
https://www.lagunaclay.com/contact/

BHOOMI POTTERY
5B Jay Chambers, Vile Parle East, Mumbai Maharashtra
400057, India
https://www.bhoomipottery.com/

CLAY STATION
#1-C, 2nd Floor, 1st "D" Main Road, 14th "B" Cross HSR
Layout, Sector 6, Bangalore 560102, India
http://claystation.in

Tools

Swivel plates and thrust bearings are available on Amazon.

Bulbs with precision tips
Xiem Tools. http://www.xiemtoolsusa.com
In India, Bhoomi Pottery sells Xiem tools. For less
expensive applicators, search for "solder flux dispenser
bottle" online.

SlabMat, for texture-free slabs, is at https://slabmat.com.

Clay Conditioner (Sodium Lignosulfonate)

VENKICHEM
A-5/201, Yogi Dham, Yogi Nagar Eksar Road, Borivali West
Mumbai 400091, India http://www.venkichem.com/

I have not as yet found a US clay supplier who stocks
ligno. The trade name is Lignotech Additive A, and the
manufacturer is:
LIGNOTECH USA INC.
100 Grand Avenue, Rothschild, WI 54474, USA
https://www.lignotech.com/

Ready-Made Tile

The supply of red bisque tile is not as consistent as white
(talc-based) bisque. If you purchase ready-made tiles,
know the temperature to which it was fired. Typically,

ready bisque is fired to 850°C–1000°C. Test-fire them a
cone higher than your glaze-firing temperature and see if
they hold up before you commit them to glaze.

Dal Tile white bisque
THE CERAMIC SHOP
1200 Markley Street
Norristown, PA 19401, USA
https://www.theceramicshop.com

Cisa-Cerdisa red Italian bisque (fired to 980°C)
NEW MEXICO CLAY
3300 Girard Boulevard NE
Albuquerque, NM 87107, USA
https://nmclay.com

Thick terracotta floor tile
NAKLANK POTTERY WORKS
8-A National Highway Road, opposite Swaminarayan
Temple, Makansar, Morbi 363642, Gujarat, India
http://naklankpottery.com

Ron Wagner